SCIENCE AND SOCIAL STRUCTURE:
A Festschrift
for Robert K. Merton

ROBERT K. MERTON

SCIENCE AND SOCIAL STRUCTURE:
A Festschrift
for Robert K. Merton

*Transactions of
The New York Academy
of Sciences*

Series II Volume 39

*The New York Academy of Sciences
New York, New York
1980*

Library of Congress Cataloging in Publication Data

Main entry under title:

Science and social structure.

(Transactions of the New York Academy of Sciences: Series 2; v. 39)
   Includes indexes.
   CONTENTS: Cournand, A. Historical details of Claude Bernard's invention of a technique for measuring the temperature and the pressure of the blood within the cavities of the heart.—Dahrendorf, R. On representative activities.—Eisenstadt, S. N. Autonomy of sociology and its emancipatory dimensions. [etc.]
   1. Science—Addresses, essays, lectures.
2. Science—Social aspects—Addresses, essays, lectures.   3. Merton, Robert King, 1910–   —Addresses, essays, lectures.   I. Merton, Robert King, 1910–
II. Gieryn, Thomas F.   III. Series: New York Academy of Sciences. Transactions: Series 2; v. 39.
Q11.N6 ser. 2, vol. 39   [Q171]   500s   [303.4'83]
ISBN 0-89766-043-9   80-13464

SP

Printed in the United States of America
ISBN 0-89766-043-9

TRANSACTIONS OF THE NEW YORK ACADEMY OF SCIENCES

SERIES II VOLUME 39

April 24, 1980

SCIENCE AND SOCIAL STRUCTURE:
A FESTSCHRIFT FOR
ROBERT K. MERTON

Editor

THOMAS F. GIERYN

———◆———

CONTENTS

PREFACE

The Board of Governors of The New York Academy of Sciences did not travel far to find the next scholar for its series honoring outstanding American scientists. It may be several hundred yards from Pupin Hall at Columbia University, where I. I. Rabi has been teaching physics since 1930, to Fayerweather Hall, where Robert K. Merton has been teaching sociology for almost as long. *A Festschrift for I. I. Rabi,* edited by Lloyd Motz, was published in November 1977 (as Volume 38, Series II of the *Transactions of the New York Academy of Sciences*), and this collection of essays dedicated to Merton is the second in the Academy's series.

The standard set by this pair will make selection of a third scientist difficult, for just as Motz could say of Rabi, "there is hardly a branch of physics (some phases of chemistry and biology may be included) that does not owe something to techniques that he developed," the same could be said of Robert Merton's influence on the social sciences. The range of subjects considered by contributors to this volume provides better evidence of this than any effusively written preface.

As it happens, the volume was almost forestalled by Merton's embarassed protests at hearing of plans for a second collection of papers to be published in his honor. *The Idea of Social Structure,* edited by Lewis Coser, had been presented to Merton just three years before, and included contributions from former students and immediate colleagues. For him, a second volume of this sort would only serve to illustrate the concept developed by Dickinson W. Richards Jr. of "hyperexis": too much of a good thing can, at times, be otherwise. Happily, the Academy ignored Merton's expressions of ambivalence, and the search was on for an editor.

Because of my ever-presence around Merton's office (as his research and teaching assistant from 1975 to 1978), I was asked to edit the volume. Surely I lacked the qualifications expected of that role. Still, I agreed, in the belief that Merton had taught me something about how to edit a book, just as he had been patiently teaching me all along about writing, addressing an audience and pursuing a scholarly career. To carry out this task would permit me to repay a small part of a personal debt for these tutorials.

The Academy and I agreed to several ground rules to guide preparation of the volume. First, there would be no overlap of contributors to this volume and those included in the Coser collection. Second, the papers would pursue topics related, however peripherally, to Merton's interest in the workings of science. Such essays, we thought, would be of greater interest to regular readers of the Academy's publications, most of them natural scientists, than essays on other sociological topics. Also, the sociology of science is among the most enduring of Merton's lines of inquiry, initiated by his 1936 dissertation at Harvard on science and technology in seventeenth century England, and gathered 36 years later in a collection of essays old and new entitled *The Sociology of Science: Theoretical and Empirical Investigations.* Moreover, the chapters in the Coser volume amply demonstrate Merton's influence on other areas of sociological research, among them structural and functional theory, role theory, the sociology of knowledge, medical education, mass communications and formal organizations.

The first rule was followed precisely; that the second was violated more than once is still another indication of the range of Merton's influence. One contributor replied to my invitation: "I shall be delighted to do it, although I am a little sorry that the subject is confined to Merton's interest in science, considering that he has had so many others." With the help of Mary Wilson Miles, who cares for some of the bureaucratic aspects of Merton's life with all of the efficiency and none of the heartlessness associated with that word, I prepared a list of scholars and scientists, other than

American sociologists, whose work I knew Merton to admire. The ultimate table of contents included three sociologists, two based in England and one in Israel. The seven Americans come from eight disciplines: economics, physiology-and-medicine, psychology, philosophy of science, information science and statistics-and-history of science. Two contributors are twice removed from Merton, in nationality and academic discipline: a historian-philosopher of science and a political scientist of science, both from Israel.

This international line-up of scholars drawn from diverse disciplines justifies Merton as a "cosmopolitan influential." Merton identified this structural type in his work with Paul Lazarsfeld on patterns of community influence during the nascent years of the Bureau of Applied Social Research. (If Merton's legacy does not provide a precise enough picture of the cosmopolitan influential, the curious reader might turn to Chapter 12 of Merton's *Social Theory and Social Structure,* recognized as a classic even by those who save that label only for the very best.) Eugene Garfield's paper provides quantitative evidence of the importance attached to Merton's ideas by scholars and scientists in disciplines other than sociology. But Garfield tells only half the story. Merton has touched so many disciplines perhaps because he has reached out to them so often for ideas. I suspect that a Garfieldian citation analysis of the sources in Merton's writings would show greater diversity in disciplinary and national origin than patterns for the typically less ecumenical sociologist. Perhaps that analysis could be part of a third volume in Merton's honor, necessitated by the fact that so many are left out of the first two who have benefitted from his insights or more directly from the bath of his red ink on a "final" draft.

Lines of influence in science and scholarship have neither beginning nor end. We always stand on someone's shoulders, and those of us on Merton's have seen farther than most. Perhaps because of the enlarged perspective provided by Merton to these authors, the essays will stimulate others to wrestle with the diverse questions they suggest. Even more, we hope that the collection offers enjoyment to the man we intend to honor, and that he will be tantalized enough to provide us all with still more fertile ideas.

Thomas F. Gieryn

Indiana University
Bloomington, Indiana

HISTORICAL DETAILS OF CLAUDE BERNARD'S INVENTION OF A TECHNIQUE FOR MEASURING THE TEMPERATURE AND THE PRESSURE OF THE BLOOD WITHIN THE CAVITIES OF THE HEART

André Cournand

Department of Medicine
College of Physicians and Surgeons
Columbia University
New York, New York 10032

The invitation tendered by The New York Academy of Sciences to contribute a paper to the volume honoring Professor Robert K. Merton gives me an opportunity to express my gratitude for the many manifestations of his friendship. Indeed, I am greatly indebted to the founder of the social science of science for sustaining, at the time of my retirement from cardio-pulmonary physiology, my efforts to focus my intellectual activities on problems related to historical and normative aspects of science.

To embark on a second career is often the aim of scientists no longer in a position to pursue their vocation. In my case, when the direction of research on the physiology and physiopathology of respiration and circulation in man was no longer possible, I was determined to maintain my close collaboration of thirty years with Dickinson W. Richards, Jr., a collaboration now oriented toward problems of science and society. I was also eager to extend my understanding of the concept and methods of the *Prospective* approach to planning and shaping the future; I had been encouraged in this by my friendly and rewarding exchanges with its promoter, the French philosopher Gaston Berger. Accordingly, I began to write on an aspect of the history of science and to prepare the text of a book on *Prospective* for readers of English.[1] It was at this juncture that Robert Merton supported my first anxious steps in my new ventures.

I surmise that Bob Merton's sympathy was probably awakened by the information—given him fortuitously—that, as a candidate for a degree in philosophy at the Sorbonne, I had been examined in 1913 by Durkheim, one of his heroes. This feeling may have been further strengthened when he learned that before undertaking an assignment to write the introductory chapter to a book published in 1962, *Circulation of the Blood: Men and Ideas,*[2] I had immersed myself for more than a year in a study of the work and methods of that great historian of science, George Sarton. The benefit that derived from Robert Merton's reading of my draft—entitled "Air and Blood"—and his unsolicited endorsement of my unexpressed hope that Sarton would have approved of my efforts, launched me on my new career. Since then, a number of papers I have written have also received the benefit of his virtually unmatched power of analysis and his immense store of knowledge.

Our relationship was also favored by our joint participation in the organization of the Institute for the Study of Science in Human Affairs, and the creation of a program on Medicine and Society at Columbia University. It was further enhanced when we became associated in the bi-yearly meetings of a colloquium held under the sponsorship of the Van Leer Foundation, the transactions of which were published under the title *Knowledge in Search of Understanding.*[3] Only in the session devoted to elabora-

1

0028–7113/80/0039–0001 $1.75/2 © 1980, NYAS

tion of the principles of the code of the scientist did I become aware of his prior interest in the subject. On that occasion, his scholarly discussion and his gracious remarks initiated a progressive momentum in the development of my thesis concerning this topic. Thereafter, his collaborator, Harriet Zuckerman, introduced me into the arcana of sociology during the many hours spent together in the preparation of a manuscript, "The code of science: Analysis and reflections on its future."[4] In maintaining as coordinates the goals of faithfully rendering Bob's thoughts and of ever-so-gently guiding me in her and my independent reflections, she displayed a grace so rare that it could have been learned, if from anyone, only from him.

As I have indicated, Bob's and my relationship has almost always found me on the receiving side. I hope that this essay will at least signal my desire to give him intellectual pleasure.

The theme of my contribution deals with the historical circumstances of Claude Bernard's invention in the middle of the last century of the technique of catheterization of the heart cavities—an invention not commonly attributed to him. Application of this technique to the study of the circulation in the human has played an important role in the modern development of cardiology. Documents brought to light in 1967 not only enable one to establish firmly Bernard's priority in the utilization of the technique described by him as "catheterism" of the cardiac cavities, their study gives detailed insights into the working habits of this great French physiologist, to whom tribute was paid this year on the hundredth anniversary of his death.[5]

We shall see that Bernard utilized both venous and arterial catheterism of the ventricles in order to resolve a physiological controversy which had vexed the best scientific minds of his day, i.e., the site of production of animal heat: the lungs or the tissues of the body? Although the data Bernard obtained in 1844 settled the question so far as he was concerned (before Liebig confirmed it using the same technique ten years later), this was not all. As techniques and instruments improved he returned to the problem, working with distinguished colleagues including Magendie, Walferdin, and Becquerel, among others. Later in his life he assembled the results of prior studies and added a comprehensive bibliography to prepare for a series of lectures given in 1872 at the Collège de France, as well as for publication of his monumental work on *Chaleur animale,* i.e., "animal heat," which was published in 1876.[6] The problem which led to the first recorded performance of cardiac catheterism was not the only one to which Bernard applied this technique of investigation. Three years later, in 1847, he recorded the blood pressure in the right ventricle.

As a preface to my detailed account of these scientific developments, I shall allude briefly to the life problems faced by Claude Bernard during the early period of his career as a physiologist. Then I shall refer to the circumstances which led to the discovery of Bernard's *Laboratory Notebooks,* in the folios of which appear the first references to the performance of cardiac catheterism. The catalogue of these notebooks and the chronological tables, including personal and scientific details of the French physiologist's life, were prepared for publication by M. D. Grmek.[7] Presently located in the library of the Collège de France, the notebooks have become an indispensable source of information in any attempt to reconstruct the history of Bernard's daily scientific activities, and to relate them to the main circumstances of his life.

CHRONOLOGY OF CLAUDE BERNARD'S LIFE FROM 1844 TO 1847

At the beginning of this period, in 1844, following the publication of his thesis on gastric juice, Claude Bernard was awarded the M.D. degree. He then served as an assistant to Magendie, who held a chair in the Collège de France. Bernard also spent a great deal of time in the chemical laboratory of Pelouze, testing the effects of various substances introduced through the venous system. In the same year he competed unsuccessfully for the title of Agrégé in the Faculté de Médecine of Paris. In December he resigned from his position as assistant to Magendie. With Lasègue, a physician who specialized in neurology and neurophysiology, he organized a private laboratory where he taught and continued his research in physiology. Plagued with

SCHEMA DES ECHANGES RESPIRATOIRES (LAVOISIER)

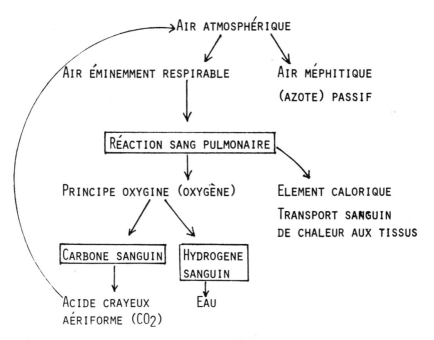

FIGURE 1. Scheme of respiratory gas exchanges in the lungs—and of liberation of the caloric element. (Lavoisier) 1. Atmospheric air is made up of "eminently respirable" and "mephitic" airs. 2. In the reaction with pulmonary blood the oxygen principle is separated from the caloric element. 3. This principle reacts with the carbon and the hydrogen of the blood to form the "chalky aeriform acid" liberated in the atmosphere, and water. 4. The caloric element born in the lungs is transported by the blood to the tissues of the body.

financial difficulties, Bernard was ready to abandon research and to establish himself as a physician in a rural district; but in July 1845 he married the daughter of a physician whose dowry (?) may have helped to solve these difficulties. By 1847 he had abandoned all idea of practicing medicine and devoted himself to advancing his research on digestion and nerve physiology. This decision became a source of conjugal dispute since his wife was a confirmed anti-vivisectionist. At the end of the year he became the acting head of Magendie's laboratory at the Collège de France.

Berger, en 1833, observait une température de 40°,90 dans le sang artériel du cœur gauche et une température supérieure, 41°,4, dans le cœur droit.

En 1844, Magendie, comme président d'une commission d'hygiène chevaline près le ministère de la guerre, avait à sa disposition un grand nombre de chevaux morveux qui, étant destinés à être abattus, étaient utilisés pour la physiologie. Nous fîmes ensemble, à cette époque, des expériences sur la température du sang dans les cavités du cœur. Voici comment j'opérais. — L'animal étant debout et vivant, je mettais à nu l'artère carotide et la veine jugulaire, et, par l'une et l'autre voie, j'introduisais un long thermomètre jusque dans le cœur. Dans ces expériences, le cœur droit se montra toujours plus chaud que le cœur gauche. Quand l'animal était à jeun la différence paraissait plus faible, elle devenait plus grande lorsque le cheval était en pleine digestion et surtout quand il venait de fournir une longue course qui avait élevé la chaleur générale du corps.

FIGURE 2.   Text of Claude Bernard's account of the first experiment performed on a horse relating to temperature differences in the blood of respectively the right and the left ventricles. Page 42. *Leçons sur la chaleur animale*, (vid. ref. 6.) (See text for English translation.)

INVENTION OF A NEW TECHNIQUE: CARDIAC CATHETERISM

The problem which was to inspire the initial performance of cardiac catheterism concerned the mechanism of production and the site of origin of heat observed in animals. The theory of Lavoisier,[8] presented in the early 1780s, held that animal heat results from combustion in the lungs (FIGURE 1). For the next fifty years the mere invocation of the great man's name was sufficient to support this belief. Then in 1837 Magnus challenged orthodoxy.[9] Having determined O_2 and CO_2 concentrations in both arterial and venous blood circulating in the same territory, he concluded that O_2 is utilized and CO_2 is formed, not in the lungs but in all the tissues of the body, and, consequently, that Lavoisier's claim was in error. Magendie and his colleague Gay-Lussac disputed Magnus's view, suggesting that his blood-gas determinations lacked reliability. It was Bernard's merit to recognize that a simple experiment would decide the issue.

The first reference to Bernard's initial experiment, performed on a horse, appears in *Chaleur animale.* The chest was closed and the animal was in a "natural state." Magendie was present as an observer. The French text of Bernard's account appears in FIGURE 2. My English translation follows:

> Magendie, as president of a committee on horse hygiene at the Department of War, had at his disposal many horses with glanders, which before being destroyed were of service in physiological experiments. . . . At that time we both cooperated in experiments on the blood pressure in cardiac cavities. This is how I proceeded: The animal being alert and standing, I exposed the right carotid artery and the jugular vein, and by both these routes I introduced a long thermometer down to the heart. In all these experiments the blood within the right ventricle always proved to be warmer than in the left. While fasting, the difference was small, but it increased at the peak of digestion or when tested after a prolonged exercise.

Concerning these experiments one may ask, how do we know that these studies represented the first use of cardiac catheterization and the first demonstration that the temperature of the blood is higher in the right than in the left ventricle? In *Chaleur animale* Claude Bernard included a table (Table B, page 48) that recapitulated the experiments of which he was aware—either his own or those reported by others in the literature. The table's first reference to cardiac catheterism as a method for measuring the temperature in both ventricles indicated that these studies had been performed in 1844. However, this table does not provide a satisfactory basis for deciding priority, inasmuch as it was constructed only in 1872, nearly thirty years after the events in question.

DISCOVERY OF CLAUDE BERNARD'S NOTEBOOKS

The publication of the *Catalogue of Notebooks* in 1967 under the auspices of the administrator of the Collège de France, Battaillon, supplied the document which allows reconstruction of the correct chronology. An account of the circumstances of the discovery of the notebooks in 1949 is given by Robert Courrier in the Preface introducing the first publication of part of these notebooks: the *Cahier de Notes: 1850–1860,*[10] known as the *Cahier Rouge.* My translation of the account of Courrier

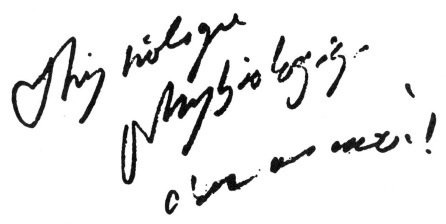

FIGURE 3. Autograph note of Claude Bernard discovered in a trunk with his notebooks. For details of the discovery by Professor Robert Courrier see text.

follows:

> On a rainy day in June 1949 I was visiting the country house situated in La Borie, in the district of Limousin (in the center of France) which Arsène d'Arsonval had willed a few years before to the Collège de France. I had been invited with Edouard Faral, administrator of the Collège de France, to attend the inauguration of a bust of the illustrious scientist born in this district, to be held in the vicinity of La Poscherie. The bad weather was no inducement to walking so I went around the house and climbed to the attic. I found there several large boxes filled with old papers, books, and notebooks. I opened some of these notebooks. In them Claude Bernard had inscribed, at the end of his laborious days, details of his experiments, projects, and fruits of lectures and reflections. As I was leafing through one of the notebooks, my eyes fell on a page where with deep emotion I read one single sentence, written with firmness of hand, in diagonal over the entire width of the paper, "Physiology, physiology, it is in me." (FIGURE 3)

RECONSTRUCTION OF THE CHRONOLOGY OF EXPERIMENTS ON THE SITE OF THE ORIGIN OF ANIMAL HEAT

The *Catalogue of Notebooks* provides a useful guide to the many experiments performed by Claude Bernard on the temperature problem. TABLE 1 lists the chronological reconstruction of those experiments.

TABLE 1

RECONSTRUCTION, IN CHRONOLOGICAL ORDER, OF THE EXPERIMENTS PERFORMED BY CLAUDE BERNARD ON ANIMAL HEAT USING THE TECHNIQUE OF CARDIAC CATHETERISM OF THE RIGHT AND LEFT VENTRICLES*

Reference	Collaborator	Date	Animal
1. "Chaleur Animale" p. 42; Table B p. 48	Magendie	Nov. 1844	Horse
2. Ms. 7G pp. 17–19 pp. 186–187		Nov. 1848 1849	Dogs (?)
3. Société de Biologie		1849	(?)
4. Ms. 8F pp. 7–13	Walferdin	May 1853 June 1853	Dog Sheep
5. Ms. 8C pp. 11–13 p. 128		Nov. 1854 March 1855	Dog Dog
6. Mss. 26A, 26C Recapitulation of experiments		March 1860	
7. Ms. 15H Recapitulation of experiments on cardiac catheterization		1877	
8. Ms. 27B Description of "Singular Disease"		July 1852	Dog
9. "Chaleur Animale" pp. 72–73 Description of thermoelectric system	Becquerel	1876	

*Problem of blood temperature before and after lung transit.

Page 15 : [*Alcali libre dans le sang.*] Y a-t-il un alcali libre dans le sang ? 1º Ajouter du chlorure de barium... (p. 15).

Pages 15-16 : *Température du foie et sucre de la veine porte.* Novembre 1848. Sur un chien adulte... (p. 15).

Page 16 : [*Section des nerfs pneumogastriques.*] Vagues coupés, absence du sucre dans le foie. Novembre 1848. Au Collège de France. Un chien loulou ayant mangé de la tripe... (p. 16).

Pages 17-19 : *Température du foie, du cœur et du poumon.* Novembre 1848. Sur un chien King Charles, adulte à jeun, j'introduis deux thermomètres très fins dans le cœur gauche et droit... (p. 17). — Le 22 novembre 1848. Au Collège de France. Sur un chien adulte... (p. 17).

Pages 16 et 18 comportent des dessins à la plume (caricatures).

Pages 19-20 et 22 : *Procédé pour découvrir le canal de Stenon chez le chien au niveau de son insertion buccale.* 1º Suivre avec le doigt... (p. 19). — Pour obtenir la salive parotidienne... (p. 20). — Analyser la salive... (p. 20).

Pages 19 et 20 comportent des dessins à la plume représentant la topographie des glandes salivaires et de leurs canaux.

Page 21 : *Pepsine.* La membrane stomacale turgide d'un chien... (p. 21).

FIGURE 4. First mention in the notebooks of an experiment utilizing right and left ventricle catheterism. (See text for English translation.)

FIGURE 4, réproducing section 7G, pages 17–19 of the *Catalogue of Notebooks,* deals with research performed in November 1848. The translation reads as follows:

On the dog King Charles, fasting, adult, I introduce two thin thermometers in left and right heart . . . (p. 17)

The note itself provides the earliest evidence of the result of an experiment demonstrating interventricular temperature differences. In this study, as I have indicated, the right ventricular temperature was higher.

Several chronologically later references to cardiac catheterism are listed in TABLE 1. Thus, there are two references to studies in which improved instruments were used. The translation of the note illustrated in FIGURE 5 (starting with "Pages 7–13") reads as follows:

Experiments on animal heat with M. Walferdin. A dog fasting for two days is killed. . . . (p. 7) A small yellow female dog at the end of digestion. . . . (p. 9) Experiments on the temperature of blood with Walferdin, 24 May 1853: 1st experiment. An adult dog. . . . (p. 10) May 25th 1853, a young female dog. . . . (p. 11) Thermometer metastatic with mercury, number 226. First rabbit-temperature between stomach and liver. . . . (p. 13)
In four folios attached to page 11 (are described) *Experiments on temperature in sheep.* Slaughterhouse of Grenelle. June 9, 1853, ambient temperature 24°, sheep. . . . June 16, 1853. Male German sheep. . . . Lesions of the left heart increase temperature. . . .

All experiments related to the use of the catheter in identifying the site of the origin of animal heat were recapitulated in the notes of 1860 and 1867. We also find in one of the notebooks (Manuscript 27b; reference 8 in TABLE 1) Bernard's description of a "singular disease" caused by the cardiac catheter in dogs. (FIGURE 6) This description, with possible side remarks interposed (a process not rare in Bernard's

Pages 7-13 : *Expériences sur la température animale avec Mr. Walferdin.*
Un chien à jeun depuis 2 jours est sacrifié... (p. 7). — Petite chienne jaune
à la fin de la digestion... (p. 9). — Expériences sur la température du sang
avec Mr. Walferdin, 24 mai 1853. 1re expérience. Un chien adulte...
(p. 10). — Le 25 mai 1853. Chienne encore jeune... (p. 11). — Thermomètre métastatique à mercure, n° 226. 1er lapin — température entre
l'estomac et le foie... (p. 13).

Quatre feuilles attachées à la page 11 : [*Expériences thermométriques sur
les moutons.*] Abattoir de Grenelle. Le 9 juin, **température ambiante 24°,
mouton**... 16 juin 1853. Moutons allemands **tous mâles**... Les lésions du
cœur gauche augmentant la température...

Pages 10-14 (partie inférieure) : *Absorption d'iodure de KO dans les
glandes salivaires.* Sur un chien de **forte taille**... (p. 10).

Pages 15-16 : *Absorption de l'acide carbonique par l'eau sucrée.* 1re expérience : dissolution du sucre de raisin... **(p. 15).**

Page 16 : *Injection d'eau sucrée dans la peau du dos.* Le 12 juin 1853.
1re expérience... (p. 16).

Pages 16-18 : *Suite de la page 28.* Dans **lequel on injecte** une égale quan-
tité de la même substance... (p. 16). — Autopsie (p. 17). — Conclusions
(p. 18).

Page 19 : *Matière colorante rouge dans les organes du bas-ventre.* 12 juin
1853. Rate de chien à jeun... (p. 19).

Pages 19-22 : *Résumé des nouvelles observations [sur la matière colorante*

FIGURE 5. Relation of experiments on dogs, rabbit, and sheep, with Walferdin, using his
metastatique thermometer. (See text for English translation.)

notes), is somewhat ambiguous, and the cause and nature of the disease ill-defined.
The English translation follows:

> The dogs die while eating, and one finds clots in the heart. These are likely emboli. There is
> no more sugar in the liver, nor in the chyle. The faculty of cellular tissue proliferations seems
> then to have stopped. ... One should reexamine the epithelium of the intestines, lungs, and
> others (to determine) if they show desquamation.
> The glycogenic cells of the liver are granulations of glycogen within the hepatic cells.
> The same granulations are present in the blastoderm of the chicken embryo.
> From the point of view of experimental pathology this is a new disease which can be
> created.

The final listing in TABLE 1 refers to the description on pages 72–73 of *Chaleur
animale* of a greatly improved method for recording temperature simultaneously in
both ventricles using thermoelectric probes designed originally by Becquerel and
Breschet in 1842.[11]

Finally, I wish to consider the admirable description of the technique of cardiac
catheterism as it is given in *Chaleur animale*.[12] Although applied to intraventricular
temperature measurement in a dog, it can still serve as a guide for the performance of
these techniques in man. The precision and the richness of details of this description
seem to me sufficient justification to provide an English translation of the paragraphs

in their entirety:

An incision is made (in the neck) to expose the jugular vein. It is by this route that right heart catheterism will proceed. For this purpose the tip of the catheter is given a slight anterior curvature, introduced into the vessel, and advanced by continuous yet not excessive pressure. Soon we feel a slight resistance. Previous experiments have shown that this occurs when the tip of the catheter has reached the atrium. The occurrence of this resistance therefore provides an important landmark. One must then rotate the probe medially and toward the left while simultaneously withdrawing it a short distance. Then, by pushing again, one enters the right ventricle. The goal has been reached.

The second step is to introduce the second catheter into the left ventricle. After exposing and isolating the internal carotid, we introduce the catheter, at the same time pulling the

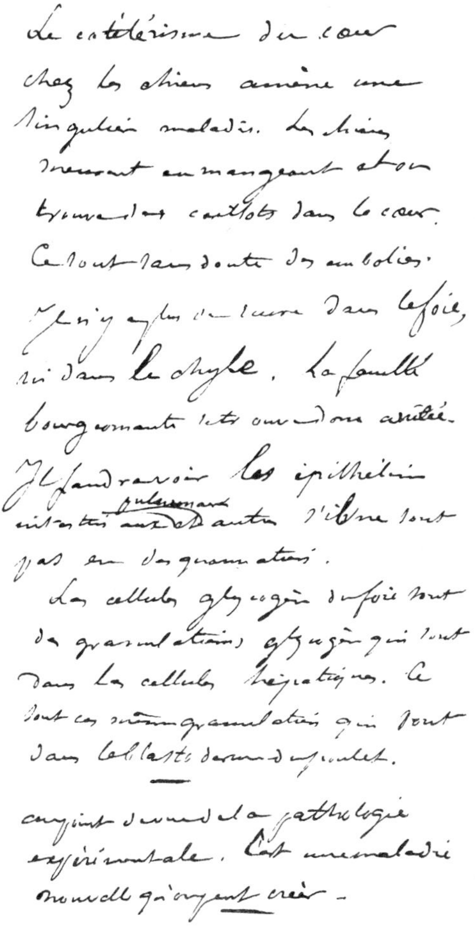

FIGURE 6. Description of a "singular disease" caused in dogs by cardiac catheterism. Besides the indication that clots of blood are present in the heart—a usual finding in any autopsy—the absence of sugar in the liver and chyle is mentioned as a possible cause. (See text for English translation.)

vessel upward in order to eliminate the curve that normally is present at the carotid artery's junction with the aorta.

Introduction of the thermoelectric probe into the left ventricle cannot always be accomplished with facility. Most of the time the sigmoid (i.e., aortic) valves prevent the passage of the catheter's tip, which tends to become lodged in the pigeon's nest formed by a leaflet and the vessel wall. One cannot advance into the ventricle without perforating or rupturing the leaflet.

After describing the anatomical and functional characteristics of the aortic valvular leaflets, Bernard goes on:

> In order to introduce the tip into the left ventricle one must wait for the moment when systole begins, as there is then a small aperture between the free edges of the valve leaflets. This is not easy. . . . However, it is easy to feel whether the tip of the catheter has entered the left ventricular cavity or is blocked at its entrance by the valve. In the latter case one sees the probe moving upward with each cardiac contraction; if one holds the proximal end of the catheter between two fingers one feels a single impact and senses that the tip of the catheter has met an obstacle and is not free. In the former case, to the contrary, one feels that the contraction is transmitted more forcefully, but that there are two distinct impacts: the first coinciding with systole and produced, without doubt, by the impact of the contracting ventricular wall on the tip, which is struck and pushed aside, and the second corresponding to diastole and probably due to the shock of closure of the aortic valve leaflets.

> In our (present) experiment, fate has favored us, and we have penetrated into the left ventricle; you can see the proximal ends of both catheters moving simultaneously, in phase with the ventricular contractions. It is necessary to secure the catheters by means of ligatures (proximal to the portals of entry into the vascular system), both to prevent hemorrhage and to avoid ejection of the tips from the cardiac cavities as a result of ventricular contractions; but the ligatures are not applied until the tips have been withdrawn slightly in order to avoid the impact of the contracting wall, as otherwise the tip would injure the endocardium or eventually perforate the muscle, resulting in intrapericardial hemorrhage.

MEASUREMENT OF BLOOD PRESSURE IN THE RIGHT VENTRICLE

We now introduce the results of a new series of experiments using right heart catheterism in order to measure the blood pressure in the right ventricle of an intact animal. In this connection Claude Bernard used the cardiodynamometer of Poiseuille, with whom he had collaborated in 1842 on studies of perfusion of the capillary system. More than 100 years earlier, in 1733, Stephen Hales had the idea of introducing a glass tube into the venous and arterial systems of a living mare in order to measure blood pressures in these systems. However, he did not extend the use of this tool to measurement of blood pressures in the heart chambers in living, closed-chest animals.[13]

Bernard was studying the physiology of the nervous system and its influence upon arterial blood pressure and for this purpose he used a cardiodynamometer. To adapt this measuring device to the proximal end of a glass tube inserted in the right ventricle involved little imagination. Retrospectively, serendipity would seem the proper word to describe this experiment, first performed in November 1847, and reported in Manuscript 7F of the notebooks.

A copy of the two folios of the complete note—obtained by courtesy of the librarian of the Collège de France—is presented in FIGURES 7 and 8. The translation follows:

> On November 12, 1847 at the Collège de France with M. Magendie and M. Rayer, on a fasting dog somewhat above middle size, a glass tube is introduced in the right ventricle

(through the exposed jugular vein) to which proximal end the instrument called the cardiodynamometer is applied. One obtains immediately the impulse of the ventricle under constant pressure. The impulse is strong, much stronger than what obtains in an artery. There is another characteristic, namely that the impulse is brusque and stiff (up and down rapidly) whereas in the artery it is undulating and slower. Is this due to a foreign body in the ventricle, or does it represent the normal state? One will have to know what is the difference in pressure between right and left ventricle. (First folio, FIGURE 7)

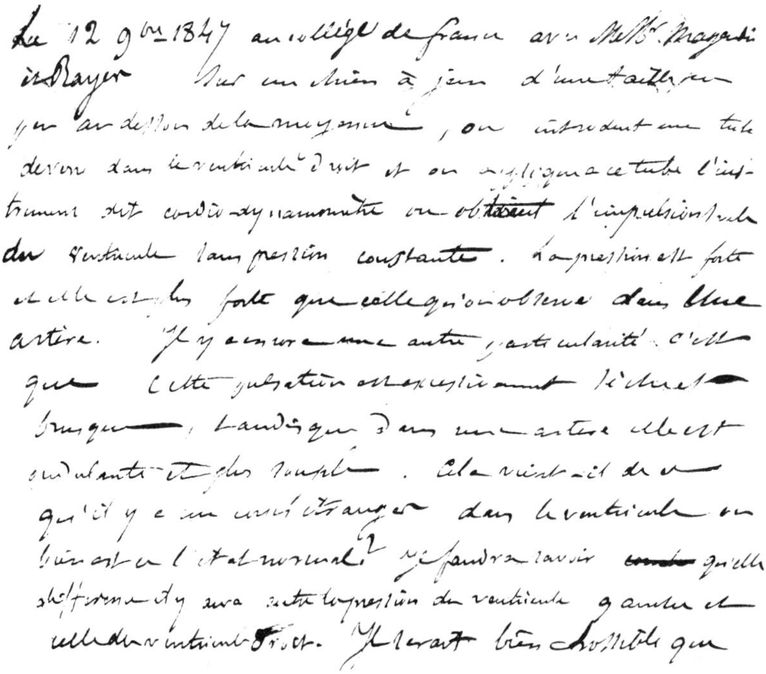

FIGURE 7. Photographic copy of the first of two folios describing the first recorded pressure of blood in the right ventricle. Ms. 7F, page 48. (See text for English translation.)

Each contraction of the ventricle gives the following data (mm Hg) from 0 to 60, 0 to 50, or 0 to 40, etc. After each peak the mercury drops to 0. There are small waves but these are difficult to identify and characterize; one notes then the following pressures:

Low	High	Average
10	72	67
20	50	30
10	45	35
0	60	10

One finds at autopsy that the ventricular wall has been perforated by the glass tube in spite of the smoothness of the distal end. There was a small amount of blood in the pericardium. However, one could hear both heart sounds. The most curious fact is the absence of constant readings in the ventricle and also in the artery. (Second folio, FIGURE 8)

Bernard continued his investigations of pressures in the ventricular cavities intermittently during at least the following seven years. His notebooks contain records

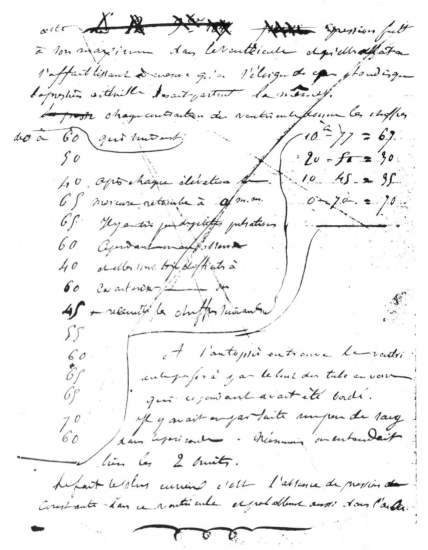

FIGURE 8. Copy of the second folio listing the first blood pressure recordings in the right ventricle of a dog and describing the autopsy findings. (See text for English translation.)

of experiments on dogs and horses in which intracardiac blood pressures were measured under a variety of physiological states and types of stimulation (notebooks 7F, 8F, 15: 1877). Later statements in the *Cahier Rouge* give evidence of Bernard's continuing interest in the problem of pressure measurement. Thus, by 1862, after publication of Chauveau and Marey's two classical reports to the Academy of Science[14a,b] on the timing of the apex beat relation to right atrial and ventricular blood pressure curves, Claude Bernard was well prepared to participate in a discussion of their findings.[15]

CONCLUDING REMARKS ON THE DEVELOPMENT OF CARDIAC CATHETERISM

The material I have reviewed supports at least two inferences, albeit with different degrees of assurance—

1. It was indeed Bernard who first performed the technique we know today as cardiac catheterization. The evidence of the notebooks as to the occasion of his first use of the method establishes his priority over Liebig. We cannot demonstrate that no one had attempted it before Bernard, but we can be sure that none of the literature available to him contained such a report—he was exemplary in his acknowledgment of the contributions of others in the problems he took up.

2. Bernard's concern for priority, at least with respect to the development and refinement of a technique, differed strikingly from that of many scientists since the seventeenth century,[16] and from that of contemporary scientists. Clearly he did not regard the invention of cardiac catheterism as meriting a separate publication. We may infer, furthermore, that directly or indirectly he conveyed the essentials of the method to Liebig, who had befriended Gay-Lussac, and to Chauveau and Marey; but the circumstances must remain hidden until further historical documents are found.

To the great French physiologist we are indebted of course for more than the invention of a method. As I indicated, he elaborated the method to a high level of sophistication. By repeating his studies with improved instrumentation and in animals of various species, and by engaging the participation of experts from other disciplines, he was able to obtain data of such quality and such import as to enhance greatly the significance of scientific study of the circulation under truly physiological conditions.

My friend Robert Merton, himself so productive in his scholarship and research, characterized a major aspect of one illustrious scientist's situation in entitling a charming work *On the Shoulders of Giants*. In relation to Claude Bernard's work on catheterism such a title would have been a misnomer. There, *he stood alone*.

REFERENCES

1. COURNAND, A. 1973. *Shaping the Future: Gaston Berger and the Concept of Prospective*. André Cournand and Maurice Lévy, Eds. Gordon & Breach Science Publishers. New York, N.Y.
2. COURNAND, A. 1964. Air and Blood. *In: Circulation of the Blood: Men and Ideas*. A. P. Fishman & D. W. Richards, Jr., Eds.: 1–70. Oxford University Press. Oxford.
3. "Knowledge in Search of Understanding." 1975. *In: The Frensham Papers*. P. A. Weiss, Ed. Futura. Mount Kisco, N.Y.
4. COURNAND, A. & H. ZUCKERMAN. 1970. The code of science: Analysis and some reflections on its future. *Studium Generale* 23(10): 941–962.
5. A Memorial Symposium: Claude Bernard and the Internal Environment. Sponsored by the Stanford University School of Medicine, Stanford, California, February 10–12, 1978. The lecture which I presented on that occasion referred partly to the material included in this essay. I am grateful to Dr. Eugene D. Robin for permission to make use of this material.
6. BERNARD, C. 1876. *Leçons sur la chaleur animale*. Librairie J. B. Baillière et Fils. Paris, France.
7. GRMEK, M. D. 1967. *Catalogue des Manuscrits de Claude Bernard*. Masson et Cie. Paris.
8. It was among Lavoisier's principal contributions to show that animal heat is produced through a chemical reaction, and not through "vital," mechanical, or other processes. His interest, of course, was in the relationship among combustion, respiration, and liberation of animal heat, and not so much in the physiological exploration of these processes; and he granted the possibility that combustion takes place elsewhere in the organism than in the lungs. For a discussion of these matters, see Hoff, H. E., Guillemin, R., and Sakiz, E.

Claude Bernard on animal heat: An unpublished manuscript and some original notes. *Perspectives in Biology and Medicine* 1965. **7**: 347–366.

9.   MAGNUS, G. 1837. Ueber die im Blute enthaltenen Gase: Sauerstoff, Stickstoff und Kohlensaure. *Ann. Phys. Chem.* **12**: 583.

10.   *Cahier de Notes: 1850–1860.* 1965. Gallimard, Paris. An annotated presentation in English of this extract of the notebooks has been published by H. Hoff, L. Guillemin, and R. Guillemin, 1967. *Cahier Rouge.* Schenkman. Cambridge, Massachusetts.

11.   BECQUEREL & BRESCHET. Mémoires sur la chaleur animale. *Annales de Sciences Naturelles, Zoologie.* Series 2, Vols. 3 and 4.

12.   BECQUEREL & BRESCHET. Mémoires sur la chaleur animale. *Annales de Sciences Naturelles, Zoologie.* Series 2, Vols. 3 and 4: 78, 80–82.

13.   HALES, S. 1964. Experiment three in *Statical Essays: Containing Haemastaticks.* Hafner. New York, N.Y.

14a.   Détermination graphique des rapports du choc du coeur avec les mouvements des oreillettes et des ventricules: Expérience faite à l'aide d'un appareil enregistreur (sphygmographe). 1861. *Compt. rend. des Séances de l'Académie des Sciences.* 7 October 1861. **53**:622.

14b.   Détermination graphique des rapports du choc du coeur avec les mouvements des oreillettes et des ventricules (deuxième note). *Compt. rend. des Séances de l'Académie des Sciences.* 6 January 1862. **54**: 32.

15.   *Compt. rend. des Séances de l'Académie des Sciences.* 28 April 1862. **54**: 899.

16.   MERTON, R. K. 1969. Behavior patterns of scientists. *Am. Scientist* **57**: 1. In this article Merton has amassed evidence that concerns about priority have been a prominent feature of scientific work since at least the seventeenth century.

ON REPRESENTATIVE ACTIVITIES

Ralf Dahrendorf

London School of Economics and Political Science
London, WC2A 2AE England

We are living in an imperfect world. Human beings are ἐνδεές, in need of things. Human needs are capable of growing, as are human faculties to satisfy them. In terms of life chances, this process may mean progress, though we do not know where, indeed whether, it ends. (We are living in a condition of uncertainty.) The mechanism of progress is conflict. From time to time it produces new things, new ways of satisfying needs. The word "produce" is unduly precise. New ways are not produced as a steel rod is produced in a rolling mill. As a result of mutations, new ways emerge. We can specify some of the conditions that make emergence possible, even probable. If we are optimistic, we assume that mankind always solves the problems which it faces; if we are not, we can argue that mankind sometimes, or often, misses its chance. But where do the ideas come from which make the emergence of new ways possible?

Why ideas? There is an enormous, and all but useless literature about the role of ideas. Vulgar Marxists seem to imply that ideas are "generated" by, or "derived" from "real conditions." That is obscure metaphoric language. The *Critique of Pure Reason* does not emerge by squeezing Prussian society. Vulgar idealists claim that ideas make history. That again is obscure metaphoric language. Weapons do not win wars; they must be used by someone, and used effectively. Using them is not necessarily a conscious process; it is like speaking prose all one's life. Nobody suggests that the first capitalist "applied" the protestant ethic.

The figure of analysis, then, is clear and simple. (Gramsci, without undue subtlety, has seen it.) There is a reservoir of ideas at any time. It includes many "relevant" and "nonrelevant" items. Social groups, or individuals, borrow from this reservoir, and thereby define relevance. Ideas become the aura, and substance, of the exercise of power. Thus there is the hegemony of ideas, their validity. By this, ideas become official, with sanctions attached to them, an order of merit for those who propound them, no promotion to senior lecturer for those who do not. There are also relevant opposition ideas, "a" rather than "non-a", likewise picked out for their usefulness. Those who propound them may not be promoted, but they are applauded by the many. The interplay of ideas and actions offers wide scope for analysis; but the question of the primacy of one or the other is quite pointless. Action without ideas is meaningless; ideas without action are irrelevant.

What ideas? This sounds as if ideas are political programs. The sound is misleading, indeed wrong. By ideas I mean all products of the human mind. (Here it would be tempting to continue: . . . which are . . . and state a number of restrictions, such as: which are original, have found a lasting form, etc.) Moreover, there is not just the world of overall sociopolitical action to which they relate. Certain areas of human activity have evolved their own action patterns: the world of science, of painting. There is, in other words, such a thing as sectoral hegemony, of relativity theory, of impressionist style. The world of ideas is a huge and in principle chaotic reservoir of possible relevance, from which given structures of power pick out elements for validity.

If ideas are a necessary condition of emergence and progress, the question remains: Where do they come from? The meaning of this question is not easy to ascertain. Nor is it easy to decide whether the question is important. Unless one assumes—we do

15

not—that ideas emerge when they are needed ("where there's a problem, there's a solution"), the field of explanation is wide open. Ideas can emerge anywhere, anytime, and it does not really matter where and when. This will not satisfy the historian/sociologist of ideas, of course; but theories of serendipity emphasize the mystery of the emergence of ideas. We will not pursue this here.

But there is one question which we have to pursue. The carriers of ideas, including new ideas, are people. We may not know, in the sense of being able to predict, who will have an idea when and how (although we know it in retrospect). It may be easier to tell under which conditions it is unlikely that anyone will have an idea, that is, there are conditions unfavorable to creativity. If scientists get no money for experimentation; if the publication of unwelcome ideas is outlawed; if a society gets so self-satisfied that no one is listening; etc.; then ideas are unlikely to come forward. (An important difference to consider, though, is: whether ideas are not generated at all, or whether they merely fail to see the light of day. The effect on action may be the same; but for understanding the history of ideas the difference is important.) However these conditions may appear, one point remains: there are obviously some people who have—who produce, generate, express—ideas in the sense in which we have used the word. Who are they? What is their specific role? What is their place in society? What does it mean for the human condition that there are such people?

In these brief and deliberately dogmatic remarks I have cut through an enormous tangle of literature under the heading of "ideology." This seems to intrigue social scientists as much as any subject, and yet much that is said is strangely barren. I have no intention of reviewing this literature here; but there may be a case for setting out briefly the kinds of problems it is intended to deal with, that is, to raise questions.

One starting point of the literature on ideology is the discovery that it is not enough to have an idea. Ideas have an effect only under certain conditions. They may be a necessary condition of effectiveness, but the sufficient condition is the state of social affairs. Do men make history? Yes, but (No, but) conditions have to be ripe, and ideas have to be there for something to happen. One of Marx's points was to remind people of this connection. Weber's "Protestant Ethic" with its methodological caution and analytical subtlety is a splendid illustration of the point.

The other starting point of the literature is related to the first, although it turns it around, as it were. Ideas have to be used to become effective—that means also that ideas are used. Ideas, and those who produce them, can become the slaves of interests. People can be bought to produce pleasing ideas, and even if they are not bought, their ideas are thus used. Much rubbish has been written about the relationship between ideas and interests. Even Mannheim's *Totaler Ideologieverdacht* raises merely the question: so what? The fact that ideas can be looked at as interest-related does not tell us anything about their truth or falsehood, value or valuelessness (cf. Merton[1]: "To consider how and how far social structures canalize the direction of scientific research is not to arraign the scientist for his motives"); and, if there is a total suspicion of ideology, we may as well discount it. While it makes sense in moral terms to consider one's interest position if one produces ideas—and while we shall argue below that the producer of ideas has a degree of responsibility for abuses of his ideas—this kind of analysis is an open-ended game for those who enjoy it and yields little for social analysis.

The third starting point of the literature is the set of questions around the conditions under which ideas are generated. This is sometimes misleadingly called "sociology of science," or "sociology of art" (or even *Soziologie der Weltanschauungen*), for it is clearly of much more general interest. It is the lifelong subject of Robert K. Merton's imaginative and insightful scholarship.

All of these approaches are related to the question of the "men of ideas" and their position, to which we shall turn presently.

There is an important passage in *The Self and the Brain* in which Karl Popper makes the point which we have in mind with characteristic fervor. I refer to it with almost total agreement. In a striking deduction (which he soaks in *Verfremdung* by using terms which are neither familiar nor likely to become familiar), Popper distinguishes the "universe of physical entities" (World 1), the "world of mental states" (World 2), and the "world of the contents of thought and, indeed, of the products of the human mind" (World 3), and says about the latter (which alone is relevant for us at this point): "By World 3, I mean the world of the products of the human mind, such as stories, explanatory myths, tools, scientific theories (whether true or false), scientific problems, social institutions, and works of art. World 3 objects are of our own making, although they are not always the result of planned production by individual men." There are important relations between the Worlds: "Many World 3 objects exist in the form of material bodies, and belong in a sense to both World 1 and World 3. Examples are sculptures, painting, and books, whether devoted to a scientific subject or to literature. A book is a physical object, and it therefore belongs to World 1; but what makes it a significant production of the human mind is its content: that which remains invariant in various copies and editions. And this content belongs to World 3." Nor is the interrelationship of Worlds merely static; World 3 has reality: "As World 3 objects, they may induce men to produce other World 3 objects and, thereby, act on World 1; and interaction with World 1—even indirect interaction—I regard as a decisive argument for calling a thing real." Or, even more powerfully put: "(1) World 3 objects are abstract . . . but nonetheless real, for they are powerful tools for changing World 1 . . . (2) World 3 objects have an effect on World 1 only through human intervention, the intervention of their makers; more especially, through being grasped, which is a World 2 process, a mental process, or more precisely, a process in which World 2 and World 3 interact. (3) We therefore have to admit that both World 3 objects and the processes of World 2 are real . . ."[2]

Who, then, lives in World 3? Before we give a positive answer, the negative answer has to be considered: not everyone. We all do things, and thus live in World 1. We all feel things, and thus live in World 2. But we do not all create scientific theories, works of art, philosophical ideas. It may be that every man has an occasional and fleeting part in World 3: understanding God, rethinking meaning, possibly producing a little poem to the beloved, a drawing, an explanation of something new. But even if this is so, it does not detract from the fact that the active and systematic life in World 3 is confined to a few. Not everyone creates new ideas.

I mention this as a fact not as an objective. Many would share Popper's scorn about "the morality of those who, being political or intellectual aristocrats, have a chance of getting into the textbooks of history."[3] There is no reason to believe that producing ideas is a "positional good" (in the sense of Fred Hirsch[4]): it is neither impossible to think of many, or even all people producing ideas, nor would it by itself detract from the value of ideas if everyone had them. Thus, why not fish in the morning, hunt in the afternoon (World 1), engage in critical criticism at night (World 3), and feel sick or happy about this life, as the case may be (World 2)?

There is a more difficult point here to which I can only allude; it has to do with quality. There is nothing wrong with encouraging creativity; the spark in every man and woman has to be kindled. But there is something wrong with a value system which no longer permits identifying quality. If the emphasis is on creativity rather than creation, on painting rather than the painting, writing rather than the book, then the emergence of ideas becomes that much more difficult, the genuinely new becomes the needle in the haystack, and fashion is likely to prevail over significance. Somehow the borderline has to be established between passive or reproductive World 3 activities, and active exploration of the open frontier of World 3. To quote Camus, "The number of bad novels must not make us forget the value of the best."[5] This requires structure, or linkage—a very precarious feature of modern societies. One of the strangest unfavorable conditions of creativity is thus the condition of universal creativity. But we leave that paradox on one side here.

Even if we allow that the distinction between Popper's first two Worlds and World 3 is not in principle one between different people—that there is no intrinsic, or *a priori* "class" structure corresponding to the Worlds—this much remains clear: no one can live in World 3 all the time. Even the monk, let alone the university professor, will occassionally *do,* and even *feel* things. More than that, most will spend the greater part of their lives in Worlds 1 and 2. There is something special about World 3, which is why it requires special attention.

World 3 activities are special in two senses: they provide a special opportunity (of exploration, innovation), and they require a special responsibility (of adding to the store of possible futures). There is something about these activities that gives them a status quite different from all others. They are carried out on behalf of people, in order to keep futures open for them. They are (as we shall say) representative activities.

The notion needs elaboration. I take it from the German philosopher Josef König.[6] The context is important. In a paper written in 1953, I had argued that freedom meant autonomous human activity, and that as such there was no difference between the dignity and significance of different activities, whether philosophical reflection, football playing, or stamp collecting. König did not like this much and replied (in a letter) with the following moving thought.

> I would so much like to add a word about "autonomy." Of course I understand it with reference to Kant, and yet—in intention at least—not in a narrow Kantian sense. It somehow pretends something of a general validity. One may regard philosophizing or, e.g., being an artist as a hobby, like playing football or collecting stamps. And at heart I always feel embarrassed to get money for it. And yet I think at the same time that it is not a hobby—but in essence more than that. There is νόμος in it, and in νόμος there seems to lie something general. Thus I find that one would not express oneself correctly, if one were to say for example, "I have made it 'my law' to go for an hour's walk in the evening" (or for "going for a walk," for example, "playing football"). It also makes no sense, or so it seems to me, to say for example: he who plays football, or collects stamps, is doing this representatively for all others. But for philosophizing, or being an artist, this strange "representative" quality might be valid on the understanding that only he does it "in the right manner" who does it "representatively." In this way one might perhaps explicate the νόμος-character in "autonomy." In this way I might perhaps even understand being paid for it, although such understanding does not remove the embarrassment of being paid. This might perhaps become the beginning of another reflection. But if I veered towards it, my thoughts would become confused. One can feel in such moments like a sailing boat which anticipates the gales into which it is running, and which sees itself drifting keel-up before this has actually happened. I believe: one must never oneself want to be certain of doing something that one is doing, representatively.

The idea is not new, of course, and its history might well be written with profit. From Aristotle to Nietzsche and beyond, it has appeared in many versions. The distinction between a βίος θεωπγπκός, a life of theory and reflection, and a βίος πρακτικός, a life of practice and action, corresponds in important respects to that between representative, or World 3 activities, and all other human activities concerning Worlds 1 and 2. The point to be emphasized above all is the need for humility; it is here that the almost tortured, deeply serious conclusion of Josef König's argument seems to me a model. We must not even think the thought of representative activities unless we include in it the notion that doing things on behalf of others is a duty and responsibility, not a privilege and distinction. Moreover, no one can say of himself he is engaged in representative activities; this is a judgement for others—and one supposes, not for contemporaries—to make.

This is not the place to begin to write the history to which I have alluded. But the point about humility is so important that there may be a case for quoting three representative authors (and thus incidentally adding to the understanding of representative activities). Aristotle clearly saw a rank order between the theoretical and the practical life. His companion distinction between work and leisure is in fact an early expression of a persistent *topos* which regards work as inferior to nonwork. There are places (in his *Politics*) where Aristotle seems to argue that the many have to work in order to give the few the chance of leisure. (This is, needless to say, what has happened in many periods of history.) In *Nicomachean Ethics,* however, Aristotle is rather more subtle and differentiates. Happiness, he says, is activity which brings out man's capacities; greatest happiness is brought by the development of man's best capacities, those of insight, of $\vartheta\epsilon\omega\rho\acute{\iota}\alpha$. It is the only activity which is not sought for the sake of something else. "We are assuming moreover that happiness must be mixed with pleasure; and the most pleasurable among the forms of activities of high value is admittedly the lively work of the philosophical mind. At any rate it holds for philosophy that it provides a pleasure which is magnificent by its purity and permanence. And it is well founded that he who operates on the basis of his insights has a more pleasurable existence than he who is only seeking the way that leads to them," to say nothing of those who are "incapable of happiness." But Aristotle adds that no one can enjoy such happiness forever except the gods who have given man but a tiny share of their pleasure.[7]

The most extreme expressions of the arrogance of representative activities can be found in Nietzsche. "A people," he says, "is the detour of nature in order to get to six, seven great men.—Yes: and in order then to get around them." Another of his aphorisms is even more relevant in our context: "Mankind does not constitute a development to the better or stronger or higher in the way in which this is believed today. 'Progress' is merely a modern idea, that is to say, a false idea ... In another sense there is the persistent success of individual cases at the most varied places of the earth and within the most varied cultures, within which indeed a higher type presents itself: something which in relation to mankind as a whole is a superman (*Übermensch*)."[8]

Albert Camus likes to quote Nietzsche, but is in fact much closer to Josef König's (and our) view, and he puts it beautifully: "One of the temptations of the artist is to believe himself solitary, and in truth he hears this shouted at him with a certain base delight. But this is not true. He stands in the midst of all, in the same rank, neither higher nor lower, with all those who are working and struggling. His very vocation, in the face of oppression, is to open the prisons and to give a voice to the sorrows and joys of all. This is where art, against its enemies, justifies itself by proving precisely that it is no one's enemy. By itself art could probably not produce the renaissance which implies justice and liberty. But without it that renaissance would be without forms and, consequently, would be nothing. Without culture, and the relative freedom it implies, society, even when perfect, is but a jungle. This is why any authentic creation is a gift to the future."[9]

Representative activities then are about producing the content of new and different things—theories, works of art, kinds of life chances. They produce the reservoir of possible futures. Without them, there can be no progress, though with them, progress is not guaranteed. Thus representative activities are both indispensable and wasteful. Whoever tries to curtail them will get stuck in the morass of existing insufficiencies, though whoever deliberately promotes them will be exposed to the discomforts of the not-yet ("*das Noch-Nicht,*" a term of Ernst Bloch's that seems preferable to the spirit of utopia.)[10] Relevant or not, there is a sense in which the creative potential of representative activities shows man at his best.

Inevitably, discussion of representative activities is reflection, consideration of the position and work of the scientist, the artist. What am I doing in writing a paper of philosophical analysis? The same is true of course for Aristotle's justification of the theoretical life, and it is hard not to suppose that Nietzsche regarded himself as one of those fortunate incidents which make whole peoples worthwhile. Other theories require other, often more elaborate constructions. If there is assumed to be an inevitable Path of History, it becomes more difficult to explain the communion of its discoverer with the World Spirit: "At times at which the class struggle approaches a decision, the process of dissolution within the ruling class, within the entire old society, takes on such an intense, glaring character, that a part of the ruling class abandons it and joins the revolutionary class . . . and especially a part of the bourgeois ideologists who have worked their way up to an understanding of the entire historical process."[11] Even Karl Mannheim must use much skill to find a place for himself in a world of "total ideological suspicion," and invent a "free-floating intelligentsia" whose members have experienced so many breaks with their primary groups that their thinking is no longer unambiguously determined by class allegiances.[12]

Those who are professionally engaged in representative activities (who get paid for them) are evidently of special interest. They include several different types even apart from consisting, by definition, of many different individuals. Some pursue *l'art pour l'art,* following an inner drive, attempting *rerum cognoscere causas,* seeking truth; there are many ways in which this has been put. Then there are the legitimizers, those who provide the material for those in power, help them do their job, explain what they are doing to others; advisers, planners, public relations people, and the like. Further, there are the rebels, the sociologists who seek what is not—not yet—institution, or not allowed to be institution. And of course there are the jesters, hard to pin down, hard to live with, the journalists perhaps, the critics from the sidelines. The distinctions are theoretical. People's motives do not necessarily coincide with their actions, let alone their effects. But the distinctions are useful as a warning against simple generalization.

There is one conspicuous omission in the catalog of representative activities which we have incidentally given so far, that is, the politician or statesman. (*Mutatis mutandis,* what we shall say applies to other leaders as well, whether in business, or organizations of any kind; but for reasons of brevity we shall concentrate on responsibility for the political community.) The omission is all the more striking since politicians—at least, though by no means only under conditions of "representative government"—are the one group which explicitly describes its activity as one on behalf of others. Nevertheless, the omission is no accident. Whatever its distinction and significance, political activity is not representative in the strict sense of World 3 activities. Politics is not undertaken on behalf of others who could and would not do it, but with respect to others who have certain more or less articulate demands; it is not representative, but (as we shall call it) legitimative.

These are subtle distinctions, and not value judgments. No implicit or explicit criterion for regarding politics or arts and sciences as more valuable is suggested here.

But we may wish to distinguish between Worlds 3a and 3b, or even introduce a World 3/2, considering Max Weber's understandable demand for "passion" as well as a "sense of proportion and responsibility" in the politician.[13] Sometimes, triteness helps: politics is the art of the possible; representative activities are the grasps for the stars. Political representatives, if they are not of the irrelevant kind that aim for nothing but high support figures in opinion polls, try to drag the indolent mass of status quo views as far ahead of itself as possible; but both moral and practical legitimacy remain crucial. They have to give reasons for what they are doing, and they have to carry the people. The good politician always has one eye on objectives and the other on support. He may strain the latter a bit if the objective is particularly important; he may also forget about the former for a while if his support is flagging. In that sense politics is impure, a balancing act which at best moves the center of the equilibrium a little forward, and at worst leaves everything as it is. There is no doubt about the creativity of politics; but it is different in quality from that of the poet or the philosopher. In Josef König's terms, one might say that there is every reason why the politician should be paid, however much one may doubt the salary of the professor. (That modern societies tend to take the opposite view is worth another reflection, taking into account how recent the phenomenon is: Bismarck still got large sums of money for his victories while Marx was not the only intellectual who was starving.) Insofar as politics is representative, this describes a straightforward relationship to a constituency which demands legitimation. Legitimation on the other hand is a term which does not arise for representative activities in our sense; they involve something that is totally removed from most people, and even of doubtful relevance, but nevertheless an expression of human possibilities which must be more than a source of personal pleasure to those engaged in it.

Much could be added to these elliptic remarks. For instance, there may be great statesmen—visionary politicians—who made "straddle" the boundary between representative and legitimative activities. (We shall return to the "straddlers" in a later note.) The place of political theorists has to be considered also. Again, the subject has a long history. Aristotle was sufficiently ambiguous on the role of the politician to allow Hannah Arendt (in her *Vita Activa*[14]) to ascribe to him many of the qualities which we have confined to representative activities here. In fact, Aristotle invented a kind of World 3/2: while politics is clearly not a form of theoretical life, it is not the practical life in the sense of the toil of the slave either; it is a distinguished mixture. On what happens, if the two vocations of science and of politics become professions as well (*Breuf*, in the double sense of the German word), Max Weber had some important things to say in the two Munich lectures of 1920. He was clearly right to single out these two human activities, whose explosive relationship accounts for so much of human history.[15]

Representative and legitimative activities are unlike all other activities of men in that their combination accounts for what may be called progress in human affairs. Representative activities create the horizon of possibilities which accompanies social life, whether as the Not-Yet or the Never. Legitimative activities shape reality by borrowing from the horizon of possibilities those items which are both desirable and capable of being implemented, given the circumstances of action. In reality, of course, this sweet division of labor is much less simple. When the great Austrian actor and theatre director Max Reinhardt died in exile in 1943, in the middle of the war, one of his eulogists said: "At a time at which political struggles demanded a clear position from the theatre, one has taken offence at the fact that the great actor had no sense for the so-called political theatre." At that time at least, it would be hard to construe any relationship to representative activities in the actions of the man who was responsible for Max Reinhardt's exile and the struggle of the times, Hitler; then, as elsewhere and at other times, staying in power has often become the sole objective of legitimative activities. Nor would it be easy to invent for the cynical and brutal politics of absolute power justifications as Max Reinhardt's eulogist could plausibly offer: "But his theatre was political in a higher sense, by being able to gather and unite what could no longer find itself on any other plane in its space of common contemplation."[16]

We are concerned here with representative activities. Legitimative activities figure merely as a mirror in which they are reflected, or perhaps, as the nonidentical twin who brings out the difference more clearly than any abstract description. For the relationship between the two activities is a privileged one. It will therefore accompany us as we look at what may be called the social structure of representative activities: the norms and sanctions that govern them; the relationship of these norms to those of society at large; and the institutions which accomodate the doll within the doll, that is, the norms of representative activities within those of the larger society.

The fact is puzzling and notable that those engaged in representative activities find themselves subject to a morality all of their own. If it is said that "this hypothesis is not testable," or "that conjecture fails to take into account Virgil's use of the word," or "he has broken the style of the introductory chapters and become unaware of his manners of expression," such judgments may have devastating effects. To be sure, the same is true if a Rotarian fails to attend 60 percent of the meetings of his club, or if a Social Democrat does not pay his party dues for a year. But there is a difference. The private rules of clubs and organizations emulate the social contract for limited ranges of social positions; they are in an important sense like social norms or laws. As a result, transgressions are explicitly sanctioned; the Rotarian is expelled, membership in the Social Democratic party lapses. In the case of representative activities on the other hand, norms are self-sanctioning in the sense that the desired result cannot be achieved unless they are complied with. This is what Josef König may have meant when he spoke of doing representative activities "in the right manner." In principle, the rules of the Rotary Club or the Social Democratic Party are arbitrary; they could be different, although in the given form they serve to constitute the organizations as they are. The same cannot be said for the rules of the game of scientific discovery or artistic invention. Scientists and artists may form clubs with Rotarian or Social Democratic rules—one cannot be a professor without having a Ph.D.; one cannot be a member of PEN without having published at least one book, and so on—but this concerns them as scientific and artistic citizens. The process of discovery and invention itself is subject to rules which are not arbitrary, even though they themselves may develop as discovery progresses. They are, in other words, an independent, separate morality.

The puzzling aspect of this morality is that it is not general. Whereas moral values like laws in principle apply to everybody, the morality of representative activities does not. If, say, a politician develops what to all intents and purposes is an untestable hypothesis in conversation with his colleagues, it does not make the slightest difference to him or anybody else whether it is testable. If a civil servant writes a paper incorporating contributions from several others, it is entirely irrelevant that the style is syncretistic, or even abominable. People are subject to the morality of representative activities only if and insofar as they engage in them; even then, they can opt out without telling anyone, or being told by anyone about it. The ethics of discovery and invention is in one sense the most voluntary of all; though for those who have chosen to submit to it, it becomes more compelling than any set of Rotarian or Social Democratic rules, or even public laws.

The phenomenon of the ethics of scientific discovery (as *pars pro toto*) requires explanation, say, in terms of the uncertainty of the human condition; we shall not pursue this line of analysis here. Given the human condition, and notably the social contract, it is not surprising that people have tried throughout history to crystallize the morality of discovery and invention and create institutions around it, academies; we shall look at an aspect of this phenomenon presently. But first it is necessary to consider briefly the relationship between the ethics of representative activities (the logic of scientific discovery, the rules of artistic invention), and the morality of social life. The issue has been given many names. Recently, "theory and practice," or, more esoterically, "praxis," has become fashionable again. Between the present and the last fashion of this terminology, "social science and value judgment" was an emotional version of the same issue. With respect to the arts, the question often arose more dramatically still, if less theoretically: what if the artist says or represents things which are shocking, objectionable, unacceptable to the moral community around him? He is censored, arrested, exiled, of course; but the underlying issue is the same conflict between the morality of representative activities and the morality of the social and political community.

Again, there is, of course, a vast literature on the subject. Perhaps a methodological remark is in place at this point. One may dispute the value of 18th century-style "think pieces" like this at a time at which most things have been said somehow somewhere, and concern with highly technical issues is the order of the day. I respect this view, though I disagree. One of the characteristics of the present state of knowledge is an abundance of information which makes it virtually impossible to see the wood for the trees. This information is valuable, indeed indispensable; it is the task of scholars and others engaged in representative activities to recognize and channel the existing stream of information. But it is an equally important and indispensable task to try and cut through the tangle (as we have put it earlier) and make sense of issues by returning to the sources. There are certain simple problems which remain such even if they are covered by seemingly endless books and articles of reflection and re-reflection. The problem of theory and practice is one of them, and we shall try to state its simple structure without reference to the literature (except for one quotation).

Marx's 11th Thesis on Feuerbach is substantially silly, politically dangerous, yet a useful starting point for the analysis of the relationship between philosophy and politics: "Philosophers have merely interpreted the world differently; what matters is to change it."[17] The thesis is false in at least three major respects; and these define the three principles that govern the relationship between representative and legitimative activities, theory and praxis.

Philosophers have merely interpreted the world differently: There is no philosopher, however conventional, limited, or second-rate, who has set out to interpret the world "differently." Incidents of forced originality, of the attempt to be different (as one suspects them in modern art sometimes) apart, philosophy, science and art are after truth. They cannot achieve it; whatever we do by way of representing the world is but an approximation, and often of melancholy imperfection; but about the attempt and its motive there can be no doubt. Far from being a cause for scorn, the differences in interpretations of the world are a testimony to the honest search for progress. Given uncertainty, that is, the constitutional inability of men to find the grail of truth, or at any rate to know that they have found it, difference and plurality are part and parcel of the moral world of discovery and invention. Where there is no difference, the values of representative activities have been violated. Marx may have dreamt of a society so perfect that it provides the real basis for a perfect superstructure of philosophy; but in the real world claims of perfection can be maintained only at the cost of suppression.

What matters is to change it: It is of course misleading to discuss one Marxian statement out of context. Marx was too much of a philosopher himself to wish to disparage philosophy as such; and when he talks of change as an almost existentialist *acte gratuite,* or in a cynical libertarian manner (Hayek: "movement for movement's sake"[18]), we must not forget that he had a notion of its direction and underlying laws of development. Nevertheless, the disjunction of philosophy and politics has to be rejected. The esoteric morality of representative activities means among other things that there is no telling what the practical effect will be if an artist, scientist, philosopher sets out on his course. There may be no effect at all; there may be a fundamental, for instance, a "Copernican" change in perspective; there may be a delayed and distorted effect like the application of Montesquieu's *Esprit de Lois* to the American constitution; there may be the kind of almost direct translation characteristic of certain areas of scientific research today. In any case, those involved in legitimate activities, in shaping reality, not only need "philosophy" in order to adumbrate their actions, but also in order to give them a sense of direction. Changing the world without an idea of direction, that is, without applying discoveries, is at best a benevolent *acte gratuite;* in fact it sounds more like the antiphilosophy of fascism, of Stalinism, and of terrorism.

The two moralities then stand in a significantly oblique relationship to each other. The values of reasonable politics require the attempt to find out what can be found out about future direction. It was unreasonable, after Keynes, to ignore the possibilities of demand management in economic policy. The values of reasonable science, on the other hand, require the attempt to disparage and supersede whatever is the established doctrine. It would have been unreasonable for economists after Keynes to breathe a sigh of relief and turn to other things; they had to wonder, when and why demand management does not work. For the politician, this is confusing. He may well be tempted to turn to Marx's thesis and discount those "philosophers" who come up with new solutions all the time. Only the most sophisticated societies understand that imposing the values of the moral community on representative activities inevitably destroys their opportunities. This is the case for the proverbial long leash on which theory must be held by practitioners, the need for patience among those who want to apply knowledge.

For those engaged in representative activities the oblique relationship between their values and those of the moral community to which they also belong, is even more painful. When Karl Popper received equally enthusiastic congratulatory addresses from conservative and socialist politicians on the occasion of his 75th birthday, he remained unworried; everyone knew his views anyway, he argued.[19] He may have been right; but the problem is more difficult. The Oppenheimer dilemma is a dramatic case in point. All modern sciences have experienced the possibility that by following their values they increase the potential of evil in the real world. Social sciences on the other hand, in their eagerness to do good, have looked more at relevance in terms of the general moral values of their societies than at compliance with the ethics of scientific discovery; as a result they have produced endless rubbish. Put in this way, the dilemma is almost incapable of solution: Ignore society, and your value-free science may lend itself to terrible abuse; embrace society, and your value-laden science will become plainly bad.

Is there an answer to the dilemma? Perhaps not, or only in an ideal world. The theoretical answer is that being engaged in representative activities involves a special responsibility, an awareness of implications. Systems of self-control with respect to biological research that may lead to genetic manipulation; or insistence by nuclear scientists that as much money must be spent on nuclear safety as on reactor construction, are practical examples of this responsibility.[20] But there is no instance that can enforce such responsibility. Scientists will do research which can be abused, and philosophers will advance ideas which can be used to legitimize evil. Politicians and other practitioners on the other hand will take whatever they can find to advance their objectives; these may be honest and good, or they may be dishonest and evil. In the process, and to make matters worse, some politicians will pretend that their policies are scientific, and some scientists will fancy themselves as having dialectically reconciled theory and practice. It is a confused world, and those in representative activities may enlighten it, but will not save it.

We have distinguished representative and legitimative activities, and discussed some of the confusions arising from the oblique relationship of their values; but increasingly there is a group of people who *straddle* the two worlds. "Philosophy" stretches out its hand in the form of "policy studies," i.e. research which accepts the subject preferences and time scales of the practitioner. "Change" on the other hand stretches its hand by adding planning departments and think tanks to government departments. (*Mutatis mutandis* the same happens in business and wherever decisions are taken.) It is doubtful whether the two hands actually meet—inquiry and action will forever follow different norms—but in the process a group emerges that feels equally at home in both worlds: scientific politicians, political scientists, advisers, planners. In most countries this is still an uneasy group which feels (rightly to some extent) that it is not recognized by either side, and whose members therefore develop ambitions either to get a university chair or a parliamentary seat. In some countries, the group of straddlers is beginning to be large enough to develop its own consciousness. This is not an easy process, even in theory, as we have seen, and it may never succeed. It is therefore not meant to be Platonic if we suggest that the kind of responsibility which we have alluded to might well be the predominant value of the straddler group: since they understand (presumably) both the requirements of scientific discovery and those of legitimate action, they can see the pitfalls in the relation of the two more clearly than others. A code of conduct for the straddlers themselves would almost by definition be an explication of the principles of responsibility with respect to the uneasy relationship of science and politics, representative and legitimative activities. There is a program here for the representatives of the Brookings Institution and the American Enterprise Institute and the Rand Corporation and others of their kind.

All institutions perish by their victories (to follow Renan), but none more rapidly and thoroughly than those designed to gather representative activities. Yet at least some and perhaps all representative activities need institutions almost as much as we all need society. "Big science" may be straddling as often as it is creative; but it needs institutes and expensive equipment in any case. Philosophers (in the widest pre-nineteenth century sense) at the very least need someone to talk to, and probably some kind of master-disciple relationship, even if the modern ideology demands calling it by other names (colleagues, *Mitarbeiter*). The "scientific community" *à la* Kuhn is both a useful fiction and an organizational reality, albeit one that rarely lives up to its moral function. The social history of art is full of instances of institutionalization, from

Maecenas to the Ford Foundation, and of course, of actual institutions, Bloomsbury, the *Bauhaus*. These institutions (I said) are almost as necessary as society. Men cannot survive without society; but representative activities ultimately transcend all boundaries of organization and institution. They are by their very nature unbounded. Whatever their institutional eggshells may be, they struggle and they fly away from them to endless horizons of truth and beauty.

The ideal-typical institution of and for representative activities is the academy, 'Ακαδεμία. This must not be confused with "academic" institutions in the modern sense, universities. It is true that in the 19th century a notion of the "idea of the university" had become fashionable which lingers on among the more abstrusely idealistic scholars of the present (or perhaps those defending an interest in remoteness?), and which claims for universities the exclusive right to house those engaged in representative activities. It is also true that universities, through the ages, have given asylum to some of them, notably in the *artium Facultates* of the Middle Ages, more generally in the new 19th century university of Europe and, later, its Anglo-American version, the Graduate School. But in all essential respects, universities have always been training institutions; a place for straddlers at best, for craftsmen as often. This is clearly true for the old faculties of theology, medicine and law; it is equally true for the mass education institutions of the late 20th century. Josef König was paid for this function of his work, not for the enjoyment of representative activity. The university is in some respects the modern, democratic version of the old aristocratic family which kept a distinguished, sometimes brilliant *Hauslehrer* to take the young brats to Italy and to the delights of Cicero and Euclid. He was bored, of course, but at least he earned a living, and even met some interesting people in the process, so that he could write his *Leviathan* or *Monadenlehre* all the more effectively.

Representative activities can be threatened by bans as well as by institutionalization. Indeed, in the modern world the choice is often between the Lubjanka and the iron cage of bondage of a bureaucratized university institute. The academy lives, under such conditions, in the interstices, in *Samizdat* circles, in structureless groups like the *Gruppe 47* of German writers, or the Bourbaki group of French mathematicians, in meeting places like the Center for Advanced Study in Palo Alto. Never has as much money been spent on the arts and sciences as in the second half of the 20th century; and yet one wonders whether there is any increase at all in the creative discoveries and inventions by representative activity. The academy is a fleeting institution with a limited lifespan; a bureaucratized world favors permanent prisons of ideas. The academy is an institution in which only ideas count, and some are clearly better than others; a democratized world favors the equality of all professors. The academy is bound to be outrageous at times and critical of established doctrine at all times; modern societies like the discomforts of doubt no more than their predecessors, even if they have found more subtle means of protecting themselves.

The figure of analysis with which we began this reflective essay was in one respect incomplete. Man is imperfect and his condition wrought with uncertainty. We are therefore seeking ourselves, or perhaps, seeking ways of improving human nature through the social conditions of life. The means by which this is done is conflict. There appears to be an eternal antagonism between what is and what could be, social institutions and social forces. Class conflicts and political struggles express this antagonism. These conflicts are not empty battles, senseless heroism like the First World War of 1914; they are about human life chances. New life chances can be discovered. Their emergence is in the first instance a matter of ideas, of discoveries and inventions. Legitimative activities need the results of representative activities to define the direction of change. Representative activities provide meaning and goals. But they provide something else as well: hope. And without hope there is no progress.

In the analysis of revolution, it has often been pointed out that a revolutionary situation is not enough. However defined, revolutionary situations are more frequent than revolutions. Nor does it make sense to assume (with Marx) that the probability, even necessity of revolution increases as the revolutionary situation becomes more extreme. On the contrary, extreme deprivation produces lethargy. What is needed to turn a revolutionary situation into revolution is the spark that sets fire to an explosive mixture of sociochemicals. This spark is hope. It consists in part in the apparent weakness of those in power; most revolutions are preceded by a slight recession of the rulers of the day. But that is merely the negative side. The positive side is some idea of what may be after the revolution, hope, in the form of an image of the future.

Revolutionary hopes are invariably disappointed. The moment of extreme freedom is soon followed by a new tyranny. Revolutionary hope is generally utopian and thus the stuff from which inhuman and unchecked power is made. Yet the figure of analysis has general application. Change requires flexible institutions, capable of adjustment without disruption. But change also requires a prospect, a design into the future, an idea of where one might go. Such ideas are provided by those engaged in representative activities. "Any authentic creation is a gift to the future." Without the emergence of new ideas by representative activities, societies become truly hopeless, drab, grey and profoundly miserable. It is useful to remember this at a time at which progress only too often seems to favor such drabness and destroy the opportunities of hope.

REFERENCES

1. MERTON, R. K. 1957. *Social Theory and Social Structure:* 532. Free Press, New York.
2. ECCLES, J. & K. R. POPPER. 1978. *The Self and the Brain:* 38, 39, 47.
3. POPPER, K. R. 1952. *The Open Society and Its Enemies.* 2nd Edition: 277. Springer, London.
4. HIRSCH, F. 1977. *Social Limits to Growth.* Routledge & Kegan Paul. London.
5. CAMUS, A. 1973. *The Myth of Sisyphus:* 92. Penguin. Harmondsworth.
6. JOSEF KÖNIG (1894–1975), Professor of Philosophy in Hamburg (1946–1954) and Göttingen (since 1954), was the author of a treatise on the concept of intuition (Der Begriff der Intuition, 1926), and one on the linguistic philosophy of being and thought (Sein und Denken, 1937).
7. ARISTOTLE, *Nicomachean Ethics.*
8. NIETZSCHE, F. *Jenseits von Gut and Böse* and *Der Antichrist* (both in: Werke III, Ed: 633, 1166. Ullstein. Schlechta, Frankfurt 1975).
9. CAMUS, A. 1973. *The Myth of Sisyphus:* 192. Penguin. Harmondsworth.
10. BLOCH, E. 1961. *Philosophische Grundfragen I—Zur Ontologie des Noch-Nicht-Seins.* Frankfurt.
11. MARX, K. & F. ENGELS. 1953. *Manifest der Kommunistischen Partei:* 19. Berlin. My translation.
12. MANNHEIM, K. 1936. *Ideology and Utopia.* Routledge & Kegan Paul. London. In Ch. 3 Mannheim quotes A. Weber as the author of the term "socially free-floating intelligentsia."
13. WEBER, M. Politik als Beruf, *In: Politische Schriften.* Weber's terms are Leidenschaft, Verantwortung, Augenmass.
14. ARENDT, H. 1960. *Vita Activa.* Piper. Munich.
15. WEBER, M. 1920. Politik als Beruf; Wissenschaft als Beruf. Both Lectures *In: Politische Schriften.* J.C.B. Mohr (Siebeck).
16. WÄLTERLIN, O. 1944. *In: In Memoriam Max Reinhardt:* 21. Zürich-New York.
17. MARX, K. *Thesen über Feuerbach.*
18. HAYEK. F. A. 1960. *The Constitution of Liberty:* 41. Routledge & Kegan Paul. London.
19. Private communication from Sir Karl Popper.
20. The Report by the Royal Commission on the Environment of 1977 (under the chairmanship of Sir Brian Flowers) provides an excellent example of such responsibility.

AUTONOMY OF SOCIOLOGY
AND ITS EMANCIPATORY DIMENSIONS

S. N. Eisenstadt

Department of Sociology
The Hebrew University of Jerusalem
Jerusalem, Israel

I

In recent discussions of the so-called crisis of sociology, one theme was strongly voiced by self-styled radical sociologists: unless sociology is true to its emancipatory vision, it is doomed to intellectual sterility and moral bankruptcy.

This claim identified the emancipatory vision with participation in the radical camp, with acceptance of its political and ideological views. It was often very closely connected with the claim that only insiders of the appropriate political, ideological or cultural camp can really understand the problems of their respective collectivities. Only "radicals" or, in the case of ethnic groups or third world societies, the radical members of these collectivities and at most their outside sympathizers, are able to uphold this emancipatory dimension of sociology.

The potentially dangerous impact of this argument on development of sociology as a scientific and scholarly discipline was forcefully analyzed by Robert K. Merton in his celebrated article on "Insiders and Outsiders."* In the following remarks, I would like to complement this argument by examining the relations between the emancipatory dimension and potential of sociology, and the development of sociology as a scholarly discipline. Or to use even an old-fashioned formulation, between the latter and the extension of the areas of freedom in public life.

II

I shall start with an examination of the nature of the emancipatory dimension of sociology. The starting point of our analysis is that, as has been pointed out in greater detail elsewhere,† sociology developed as a part of the broader intellectual tradition of self-examination and self-inquiry that developed in Europe in the wane of the Reformation and Enlightenment. It focused on the ciritical approach to the basic phenomena of human and social existence and to the nature and working of social orders.

Within this broad tradition, sociology has been distinguished from the philosophical, political-ideological and reformatory trends by an emphasis on the critical, scholarly approach to the analysis of social life. It was within the framework of its development in relation to these broad traditions that the special emancipatory potential of sociology has been developed. This potential has been manifest in the broadening of the self-critical attitude in general, and of the critical and enlightened

---

*See R. K. Merton. 1972. Insiders and Outsiders: A Chapter in the Sociology of Knowledge. *The American Journal of Sociology,* **28** (1): 9–42.

†See on this in greater detail. S. N. Eisenstadt and M. Curelaru. 1976. *The Form of Sociology: Paradigm & Crises.* John Wiley & Sons, New York, N.Y.

0028–7113/80/0039–0028 $1.75/2 © 1980, NYAS

analysis of social life in particular. In taking this attitude our of purely philosophical or political and ideological frameworks, sociology has created a distinctively critical yet scholarly orientation to social life.

The emancipatory potential of sociology could be maximized by subjecting problems of social life—including political, ideological or reformist assumptions—to such critical scholarly analysis. By working in this way, sociology could broaden the self-critical attitude in modern societies, or enlighten public debate. To dissociate these problems to some extent from direct political and ideological involvements and pressures extends the range of so-called reflexivity in society and in this way contributes to the extension of the range of freedom in society.

Needless to say, areas of sociological research and analysis were differently affected by these issues. It was but natural that the most critical testing ground of these approaches was neither methodology nor abstract theory, but those middle range theories which address themselves to the analysis of contemporary social reality in its diverse aspects.

But, significantly enough, it was only in so far as such middle range theories and analyses were closely connected with other fields of sociological analysis, i.e. related to broader theoretical concerns and based on social science methodology, that they could go beyond mere ad hoc descriptive and journalistic attempts and have their specific contribution. The ability of sociologists to work in such ways was largely dependent, above all, on the degree to which the sociological community could maintain its own intellectual and institutional autonomy—above all to develop institutional settings in which the different fields of research could be brought together. The importance of these conditions became greater the more sociology was practiced, not only by gifted individual scholars, but as it become institutionalized in academic life and in research institutes.

III

But such autonomy and the concomitant distinctiveness of the sociological tradition were never automatically assured. There are several reasons for this. First, sociology never became totally separated from other parts of the broader intellectual context of which it was a part. Indeed, these close connections could be found in most strands of sociological thought, and many of the major developments were closely related to various "external" intellectual (e.g. philosophical, political and ideological) and social reformist traditions. Sociology was never closed to these external influences, and these have in some cases led to the weakening of the distinctiveness and autonomy of the sociological tradition. Second, this autonomy could be undermined by the nature of the institutionalization of the discipline. Institutionalization could lead to the fractionalization of different types of social research. For example, it might lead to the growing involvement of some sociologists in public and private agencies, with the resulting acceptance of the more technocratic orientations often found in such agencies.

Have these influences broadened the sociological tradition through the reformulation of problems of research according to autonomous critical scholarly criteria, or have they taken place in such a way as to lead to the abdication of the distinctiveness and autonomy of sociological analysis?

Such abdication developed in several directions. One has been the abdication of the distinctiveness and autonomy of sociology in favor of philosophical or mostly ideological and political tenets, dogmas and pronouncements. The second direction of abdication has been the concentration by sociologists on purely technical aspects of sociological analysis—often combined with technocratic tendencies and the consequent closure of sociological work to broader concerns. Both these tendencies were connected with or reinforced by developments and trends in the organization of the sociological community—especially by the development of sectarian tendencies on the one hand and of technocratic orientations on the other. Both types of abdications have usually involved a certain selection from within the various types of sociological research and analysis. Thus the tendency to abdicate sociology "in favor" of ideological and political positions usually entailed a concentration on philosophical, epistemological-methodological discourses, while the abdication in a more technocratic direction has usually entailed the heavy, almost exclusive concentration on the purely methodological or survey type of research. Common to both types was the withdrawal from that type of research most closely related to the emancipatory dimension of sociology: middle range theories concerned with the analysis of the workings of societies and of contemporary social reality in its diverse aspects.

IV

The selective emphasis on different aspects of sociological research and on the emancipatory dimension of sociology can be seen most closely in those cases where such abdication was imposed on the sociological community entirely from the outside, such as cases of political repression.

In most autocratic, but especially in totalitarian regimes (communist ones in general, the Russian case in particular), sociology was allowed to develop only in its methodological aspects (e.g. the technique of survey research and the statistical manipulation of data) and in its highly abstract theoretical aspects. That component of the sociological tradition which focused on the analytical-critical study of society—based not on dogma but on scholarly reflection—were discouraged, and where it did develop, it often served as the butt of political censorship.

Through their selective adaptation of certain aspects of the sociological tradition, the totalitarian regimes indicate their good understanding of the potentially emancipatory functions of some sociology. By avoiding theories of the middle range, which provide a critical analysis of societies such as their own, they have identified those aspects of the discipline which are most threatening to totalitarianism. Sociology in "open" societies has largely been able to overcome these abdicative tendencies, and to maintain a scholarly and emancipatory momentum.

V

One paradoxical aspect of the debates surrounding the potential for an autonomous sociology is that these debates themselves may in fact contribute to a move away from the emancipatory dimensions of the discipline, not through political oppression but through internal developments of the field. Groups of self-styled radicals advocate the identification of sociology with a certain set of ideological premises and the use of that ideology as criteria for the formulation and evaluation of sociological problems. This position is often accompanied by the view that *these* efforts represent the epitome of emancipatory sociology. In fact, such positions proved antithetical to the development of the emancipatory potential of sociology and to its intellectual progress.

These radical views fed on institutional developments within sociology during the 1950s. There were tensions emerging from the opposition of enlarged numbers of students in sociology and an outmoded organization of sociological activities. The gap between the pretensions of some sociologists to solve the ills of the world and their modest successes reinforced the tendencies. Also, the strong technocratic tendencies and the growing specialization of social research may have contributed to the desire by radicals to solve the problem of sociology by combining it with politics and ideology.

These developments may undermine the attainment of several constructive conditions: first, the ability of sociologists to form a collective solidarity and an identity that would encourage an openness to broader intellectual developments without sacrificing the distinctiveness of sociology; second, a commitment to objective exploration and, at the same time, a commitment to a critical orientation to societal and political problems; third, an encompassing framework for the discipline which would provide some organization for the diverse, specialized activiies of sociologists.

That it was possible for sociological communities to forge some measure of consensus and cohesion is one of the paradoxes in the development of the field. The most important challenge before sociologists today is to maximize (very much in the spirit of Merton's life-work) the constructive potential in an autonomous sociology, while minimizing the often concomitant sectarian and ideological tendencies.

OF CUNNING REASON

Yehuda Elkana

The Van Leer Jerusalem Foundation
The Hebrew University of Jerusalem

GROWTH OF KNOWLEDGE

The historical sociology of scientific knowledge is an emerging area of study, deriving its intellectual energy from several sources, the work of Robert K. Merton being of the most important among them.

The problem of knowledge is a problem of change. What we need is a theory of the growth of knowledge which would account for development and change in our culture, in other cultures, and in the individual, and which would satisfactorily explain cultural differentials as well as point to cultural universals. The field dealing with such a problem is a combined history of science, sociology of knowledge, anthropology and genetic epistemology, always with strong emphasis on the comparative. At present, several different fields exist. Genetic epistemology and structural linguistics, or rather, generative semantics deal with the growth of knowledge in the individual. Anthropology, as it is today, emphasizes the differentiae: it is relativistic with respect to ethics and religion but embodies a positivistic view of science and, therefore, while laying stress on comparative religion, comparative magic and generally comparative ethics, there is no comparative science. Sociology of knowledge, the little that has survived the Mannheimian failure, generally goes hand in hand with anthropology, although it deals more with Western cultures and their development; yet it shares with most anthropology the combination of an antipositivistic bias in the social sciences with a positivistic view of science. Thus it turns out that in an age when science is considered the most important human enterprise and the greatest positive achievement of the human mind, the prevailing conception of it eliminates any attempt of dealing with its development—namely an attempt of giving a theory of the growth of knowledge—in most of the fields which should be dealing with science. Only the history of science, or rather that recent amalgamation of the history, philosophy and sociology of science which Merton has called Historical Sociology of Scientific Knowledge, has become the framework for theories of the growth of knowledge. What is meant by knowledge is generally scientific knowledge, science being considered the rational enterprise, par excellence, and thus it turns out that the search for a theory of the growth of knowledge becomes the problem of rationality of scientific change.

The problem of understanding other cultures (the anthropological problem), of understanding past periods in our own culture (the problem of history of science), and the stages in, or process of, growth of knowledge in the individual (cognitive or Piagetian studies, moral or Kohlbergian studies and Eriksonian stages) are all essentially the same problem.

Every culture and every individual at any age have a view of the natural world, of society and of knowledge—be it conscious or not, well articulated or not, in accordance with what we, in our Western culture accept as valid knowledge or not. All learning, i.e. growth of knowledge, is the replacement of previous views by newer, more acceptable ones and never is new information filling a void in the intellectual space.

As a result of this presupposition, it will be argued that all cultures as well as all individuals think consciously at least about some of these problems some of the time and thus exhibit rationality to some degree: they have some kind of science.

0028–7113/80/0039–0032 $1.75/2 © 1980, NYAS

This view is relativistic in terms of different cultural frameworks but not relativistic inside a framework. That is: since we cannot, and will never, be able to learn all cultural frameworks and elevate ourselves to an extraterrestrial point of view which could possibly find Truth from an absolute cultural inertial frame of reference, there is no absolute way of comparing cultures on a single scale. However, there is no reason why we, in our own framework, should not develop criteria for the best theory, the best result, the best method without claiming an eternal, culture-independent or even a time-independent status for these presuppositions. What I am describing here is what I call elsewhere two-tier thinking.[1] That we are capable of it is seen from the very fact that we do learn from our mistakes, that concepts change and theories develop; each change and development is a quick moving shuttle between different frameworks of presuppositions. It is the shifting sands of changing presuppositions, be it ideas in the body of knowledge or views about knowledge (that is, images of knowledge) that we wish to explain rationally. Such a view will make most demarcation criteria between science and nonscience, rationality and irrationality, superfluous.

The growth of knowledge is a result of continuous critical dialogue between competing scientific metaphysics, competing images of knowledge and competing normative views. All three are part of comprehensive scientific research programs.

The body of knowledge (with scientific metaphysics at its centre) is the sum total of statements about nature or man or society (or any of its special branches); the images of knowledge are views of knowledge, its sources and its validity. It is socially determined and varies from culture to culture, from period to period, from stage of development to stage of development. The third factor is the normative embodiment of sociopolitical forces. They all interact in a complex historical way. This makes any attempt at a permanent, absolute distinction between internal and external history artificial and superfluous.

The "thinking" in the body of knowledge is a combination of paradigmatic and syntagmatic thinking. Sometimes we try to think by analogy to paradigmatic cases and at other times to subsume the phenomena to known laws. Both types of thinking occur simultaneously.

Science is a cultural system like ideology or religion or commonsense (echoes of Geertz), and just like in religion, one can always distinguish in a tradition between the great, abstract theories and their local version. Science, too, has a Great Tradition and a Little Tradition.

A theory of the growth of knowledge, through a critical dialogue between the competing theories and metaphysics in the body of knowledge and between the competing images of knowledge (source of knowledge, aims of knowledge), is a theory which coheres with the Historical Sociology of Scientific Knowledge. It replaces the accumulation view, the revolutionary view and the various evolutionary theories of growth of knowledge.* It allows for a distinction (albeit temporary) between body of knowledge and image of knowledge, but abolishes the demarcation between science and nonscience, between external and internal factors, and between the context of justification and the context of discovery. It rejects the deterministic view of the emergence of knowledge and it rejects the primacy of epistemic reason. In short, it rejects the view that by way of "rational reconstruction" some positive developments can be explained (having given for their occurrence both necessary and sufficient conditions), while great historical changes are not explained but rather ascribed to a vague change in Zeitgeist or Weltanschauung. Thus the programme of the Historical

---

*The major problem of the evolutionary theories in my opinion is their attempt to view ideas as analogues of mutants.

Sociology of Scientific Knowledge is to give at least the necessary conditions for all historical events without presupposing that what happened had to happen.

SCIENCE, THE EPIC THEATRE

There are two alternative approaches to history. One is the perspective of Greek drama, the other is the perspective of the epic theatre. Since theatre, good theatre, is indeed a mirror of all that there is, an analysis of these two world-views will lend us the two general perspectives for viewing history.

Greek drama is a development of the inevitable. Fate is immutable, and man can influence only in minor details the when and where of his own destiny. The very tension that we experience in drama is caused by our knowledge of the inevitable. It is only our sense of the pending doom that makes us fear what is coming, thus creating dramatic tension. The Greek tragic view of history is that the future will unfold according to pre-established rules and once an event has occurred, it becomes clear that this is the only way it could have possibly occurred.

It is an old Western cultural tradition to view the growth of knowledge— knowledge of all kinds, even scientific knowledge—as the subject of Greek drama: the unfolding of the inevitable. Already Epicurus realized this and feared the straight-jacket of science, saying:

> It would be better to follow the myths about the gods, than to become a slave to the physicist's destiny. Myths tell us that we can hope to soften the gods' hearts by worshipping them whereas destiny involves implacable necessity.[2]

Such a view of science leads us to believe that there is only our science to be discovered, that the great truths of nature, had they not been discovered by a Newton or an Einstein, would have been discovered by someone else sooner or later; that unlike religion, or art, or music, or political ideology, there is no such thing as "comparative science" among different cultures, and any attempt at creating it is meaningless; that the temporal unfolding of our knowledge of the world, though it may be an infinite process, yet is fixed and inevitable.

Not so in epic theatre. The idea of epic theatre has been developed by Walter Benjamin and by Bertold Brecht and its main historical thesis, as formulated by Benjamin, is very simple but in glaring contrast to Greek drama: "It can happen this way, but it can also happen quite a different way."[3] Thus, the historical question is not "what were the sufficient and necessary conditions for an event that took place" but rather "what were the necessary conditions for the way things happened, although they could have happened otherwise."

Such a view is a very undramatic one. It is typical of the Chinese theatre to "make what is shown on the stage unsensational," but it is also an antithema in Greek culture. As Plato had already recognized, and it has been repeatedly claimed ever since, the highest form of man, the *sage,* is of a very undramatic nature.† The sage is not "a Stoic figure who has learned to overcome his involvement in the 'momentary and the personal' (in Einstein's words), but one who has eliminated it altogether, and exists on a different level."

Epic thinkers are no Stoics, and they do not hold the Greek tragic view of science. The world is an epic theatre: the world of Reality happened to have taken place in the way it did. The future is always unpredictable and any event that had occurred could

---

†Many of the Brechtian 'heroes' are such, as pointed out by Benjamin: Galy Gay in "A Man is a Man" and Azdak in "The Caucasian Chalk Circle."

have happened differently (had reality been different). The same in the moral sphere. Unlike Seneca, according to whom "the Divinity has determined all things by an inexorable law of destiny, which he had decreed, but which he himself obeys,"[4] epic thinking opts for a freely chosen moral code, which is analogous to the great unifying theories of the world. There the symmetry, the fundamental simplicity, the rational 'inner perfection' created an epic distance from the dramatic and from the 'merely personal.'

In epic theatre the only historically meaningful question is: why did it happen the way it did? It could have happened otherwise. On this view science could have been developed differently, other discoverers could have discovered different laws of nature. There is nothing inevitable in the uniqueness of Western science; a 'comparative science' between different cultures is meaningful; lessons can be drawn from history for future use. All in all, this is an optimistic perspective.

All *idealistic attitudes,*‡ whether reductionism, positivism or behaviorism, share the Greek drama view of science. On the other hand, I fully identify with the historical attitude of the epic theatre, and it is this world-view which underlies the discussion here.

MODES OF REASONING: CUNNING VS. EPISTEMIC REASON

Alongside with the deterministic Greek drama conception of knowledge and especially of scientific knowledge, fittingly, an exclusive reliance on epistemic reason has accompanied the development of Western culture. Reliance on logic and rationality, objectivity and the impartiality of the scientific quest, implies a distinction between the treatment of the external world and the interaction with human individuals or groups. It is presupposed that the 'world' does not react to our interference with it. Though we may pose questions to Nature, yet its answers will not depend on who posed the question or how the question was formulated. Even in sociology and in some psychology (behaviorism, psychophysics), the derived end result of research is to objectivize the process, so to formulate the problems that the solution would be independent of the people and processes involved. This is typical of epistemic reasoning. But there is another kind of reasoning, too. This is cunning reason—the same cunning reason that Einstein attributes to God in his famous dictum "God is cunning but he is not malicious," or Bacon to nature when he says that nature's secrets must be "wormed out of her." This type of reasoning and of knowledge has always been with mankind, but it was generally overshadowed by the official epistemic reason which is typical of logic and of all rational reconstruction. Already in the Greek sources alongside *episteme,* the cunning reason, *metis,* can be seen in action, only to be dismissed as merely practical and lacking rigour. Yet it plays an important role in Homer, in Greek drama, and even in Plato.

> *Metis* implies a complex but very coherent body of mental attitudes and intellectual behavior which combines flair, wisdom, forethought, subtely of mind, deception, resourcefulness, vigilance, opportunism, various skills and experience acquired over the years. It is applied to situations which are transient, shifting, disconcerting and ambiguous, situations which do not lend themselves to precise measurements, exact calculation or vigorous logic.[5]

The rational reconstructionist would like to disregard cunning reason, but it is an

‡'Idealistic' is italicized to emphasize the somewhat idiosyncratic though not unintentional use of the term.

integral part of the scientific enterprise. When applying cunning reason, in addition to what we want to say or ask, we also must consider the influence it will have on the dialogue partner, and we must tailor our formulation to our partner. We act so (consciously or otherwise) in daily contact with people: we convince, cajole, manipulate, charm, disappoint, enrage other people. We do so also in areas where our enterprise is scientific: like in anthropology or psychology and moreover that cunning reason is what typifies the process of discovery even in the natural sciences.

In anthropology, cunning reason is applied in *thick description*.[6] Thick description is the interactionist view, two-tier-thinking, and back-and-forth movement between sense experiences and speculation. It escapes the artificial choice between 'total' objectivity and 'complete' relativism.

The term "thick description" is Ryle's. "Thick description" is the most fundamental everyday activity of the ethnographer. For Ryle, "thick description" is a way of describing the complexity of thinking: he starts from the most elementary one-layer activity, like for example, counting the number of cars on the street. Describing this activity involves a very "thin description." Then, layer by layer (or step-by-step on a ladder) the activity becomes more complex and its description thicker. Thus, the kind of description we have to give when describing what a person is doing is somewhere on a continuum between the very thin to the very thick, and the thickness depends on the kind of activity we are describing. In ethnography the problem is always one of translation (from culture to culture), so whatever the ethnographer is describing can no longer be a thin description.

All this multiplicity of complex conceptual structures, many of them superimposed onto one another, is also a good description of what a scientist is doing: formulating problems, choosing phenomena, demarcating the seemingly self-evident from the seemingly puzzling, observing selected relevant motions, changes, processes, sizes, color, etc., reducing one phenomenon to another, and then changing frameworks by going vice versa, describing his experiments. No doubt he has to deal with a multiplicity of complex conceptual structures, mostly superimposed on one another.

This is actually a manipulation of nature. A complex interaction of layer-upon-layer observations, speculations, confirmation and falsifications creates a dialogue between the researcher and his object. What he will do next depends greatly on what he interprets, and the way he formulates his tentative probes will greatly influence his own interpretation of the resulting facts. All this is cunning reason.

An interesting example of the application of metis occurs in legal procedure. The Chief Justice of the Supreme Court appoints three judges to hear a case and instructs one of them to write a summary. In order to become a ruling, at least one of the two others has to accept the summary of the one who wrote it. If the two others reject it, the summary remains an empty argument. Thus not only does the judge who was instructed to draft the court's summary have to stick to his views, but he must take into account the character, biases and legal history of his colleagues in order to formulate his summary so as to make it acceptable to the others or at least to one of them. He has to apply cunning reason.§

The Context of Justification and the Context of Discovery

Most of the best works in history and philosophy of science in the last decades were aimed at studying how much of a role did tradition, ratiocination, intuition, values and

§I owe this example to a friend.

authority play in the acceptance and rejection of scientific theories and in addition to observation and experimentation. It is often objected that, since these considerations relate only to the acceptance (or rejection) of theories, they are of relevance only in the context of discovery and not in the context of justification. In my view they are relevant as much for the context of discovery as for the context of justification, even if we think that these two contexts can be distinguished. Tradition, authority and values play as much a role in creating tools for justification as in telling the story of a discovery. However, the situation is more complex than that. The distinction between context of justification and context of discovery was introduced by Hans Reichenbach[7] and it referred mainly to mathematical discovery. It was quickly accepted by the Reichenbach circle and by the logical positivists (by then mostly in America), and it became strongly connected with the observation-theory distinction, with the emphasis on the internal vs. external causes for changes in scientific knowledge. In short, it became part and parcel of the so-called "Received View."[8] For them, the genesis of theories is not interesting, or at least, it has no epistemological relevance. The importance lies in dealing with the finished products of science, i.e. in their justification (or in Lakatosian terms, in the rational reconstruction of the theories). At the same time, for Wittgenstein (especially in the *Philosophical Investigations*) and for his followers, the context of discovery became more and more important. However, neither Wittgenstein nor his followers, like Hanson or Toulmin, get down to the socially defined part of such Weltanschauungen. Suppe, when giving a summary of views alternative to the "Received View," shows full awareness of this problem.

> Such a Weltanschauungen approach to analyzing the epistemology of science obviously must pay considerable attention to the history of science and the sociological factors influencing the development, articulation, employment and acceptance or rejection of Weltanschauungen in science. As such, the concerns of the philosophers of science overlap those of the historian and the sociologist of science.[9]

Mary Douglas,[10] in the preface to a recent collection of essays, radicalizes and continues Durkheim's relativism: the knowledge of the universe is socially constructed, and this is the Durkheimian theory of the sacred. Durkheim has shown the self-delusion in imagining non-context-dependent, culture free, 'objective' religious truth, and this turned all the religious establishment against him. Durkheim, like Helmholtz, in the Preface to his famous 1847 paper on "The Conservation of Force,"[11] like the psychophysicists Weber and Fechner, like Emile Meyerson[12] in his *Identity and Reality* (actually, like most neo-Kantians) also explained how this self-delusion came about. As Mary Douglas put it:

> Delusion is necessary. For unless the sacred beings are credited with autonomous existence, their coercive power is weakened and with it the fragile social agreement which gave them being. A good part of the human predicament is always to be unaware of the mind's own generative powers and to be limited by concepts of the mind's own fashioning.[13]

Although Durkheim tacitly implied it, he did not claim the same for secular knowledge. He did not even try to show the delusion involved in the cultural fixation on the 'sacred' of secular knowledge, namely on 'scientific truth.' Scientific truth, for Durkheim, remained genuinely irreducible. Mary Douglas radicalizes Durkheim, brings his theory into coherence with Wittgenstein, and firmly announces in their name that:

> even the truth of mathematics, let alone other truths are established by social process and protected by convention.[14]

The problem is shifting, however. Can the context of justification be neatly distinguished from the context of discovery? Is it always clear to us in what activity we are engaged? The question is rhetorical and the answer is negative both for doing science proper and for discussing science, be it historically, philosophically or sociologically. The first is easier to show: it is well known how many cases of discovery happened when trying to corroborate or falsify by argument or by experiment a previous result. This can be formulated even in a sharper form: every experiment is a multiple choice test between a few (generally two) possibilities which remained after a previous elimination according to a predetermined theoretical context. All experimental discoveries happen much in the same manner; this applies even for cases where a discovery is genuinely experimental, and not theoretically anticipated. In most cases, however, the theoretical discovery precedes the experiment; this mostly occurs while trying to argue out or falsify (i.e. to justify) previous theoretical results. It is somewhat more complex for metatheoretical discussions of science or for grand new world-views, i.e. scientific metaphysics. Even there, I believe, the two contexts are intertwined, since all new ideas (whether theories, world-views, metaphysics, images of knowledge) occur as replacements of previous ideas and never in a vacuum, and are influenced by the replaced old ideas.

The distinction between justification and discovery is mostly used for describing the activity of historians and philosophers of science, not for the activity of scientists. To say that here, too, the distinction is spurious, is to say that a theory of growth of knowledge has already been espoused with the accompanying images of knowledge. If science is identified with the artificially delineated 'finished products' which are obtained by fixing disciplines or subdisciplines and eliminating the open problems from them—in short, by mistakenly taking the 'textbook' for the cognitive field—then justification is not discovery. If one holds that all reasons for acceptance or rejection, corroboration or falsification of specific results, stand alone, irrespective of the acceptance or rejection of other parts of the theory, then again justification is not discovery.** Furthermore, if it is believed that all reasons for acceptance or rejection are to be found in the body of knowledge (and not in our views about knowledge, that is, in the images of knowledge) then once more the distinction may hold. If, however, it is held, as I do, that knowledge grows by a continuous critical dialogue between competing scientific research programs with their competing scientific metaphysics and competing images of knowledge, then in the flux there are no finished products, there is no elimination of the unknown, no tight separation of open problems from the solved ones and consequently, no distinction between the two contexts. We have seen that whosoever talks of the context of justification is referring to epistemology or to epistemic reasoning. Philosophers or scientists who think exclusively in terms of episteme cannot explain rationally a discovery. There is no logic of discovery and thus there is no account of discovery.

If, on the other hand, we do not accept a demarcation between the two contexts, and we try to describe the process of discovery in terms of thick description and images of knowledge (that is, in social terms), then we are again in the realm of cunning reason describing the interaction between interrogator and subject of inquiry. Thus if a distinction between the two contexts is admitted at all, then the context of justification will argue in the framework of epistemic reason, while the context of discovery in that of metic reason.

_____

**Very rare nowadays: the Popperians as much as Mary Hesse or Quine or Putnam speak of networks of theories which stand or fall together.

SOME EXAMPLES

The Mind Is No Waxen Tablet

Francis Bacon said:

> On waxen tablets you cannot write anything new until you rub out the old. With the mind it is not so: there you cannot rub out the old till you have written in the new.[5]

What Bacon rightly says is that the prevalent view of the process of education, namely that knowledge is being transferred from a full bottle into an empty bottle, is plain wrong. Understanding a problem or a question means already having a spectrum of possible answers, that is, an elimination of many logically possible solutions. There is no such situation that the student, whatever his age, does not have already a view on the issue discussed. The teacher's task is only to prove or argue that his version, information, theory is the correct or at least the better one, and that it should replace the previous one. And with the mind—the old will disappear only after the new has been written in it.

Much follows from this. Before any new theory is introduced, an attempt has to be made to find out what each student happens to think (rightly or wrongly) on the question raised. (Needless to say, this is practically impossible. However, from time to time it can be shown that the very understanding of a question involves some answers, and that about every new theory or issue there are some prejudged views and attitudes.) The rest can be done by using historical case studies. To research in depth the views, scientific metaphysics and images of knowledge of an individual or a group or community prior to, and consequent to, a scientific discovery can be an eye-opener. This incidentally can serve as a guide to the great variety of ways of discovery. There is no one method of discovery nor a logic of discovery; there is only a plurality of open possibilities. In order to find out what are the preconceived ideas to be replaced involves *cunning reason*. The very question we formulate will determine, at least in part, what we find as a possible answer.

Is the Meaning in the Text?

One allegedly distinctive feature of science is the sufficiency of the text. The attribution of meaning in science, in contrast to religion or art, is supposed to be independent of personal idiosyncracy and cultural background. This view is closely connected with the theory that scientific knowledge accumulates over diverse epochs and various places, and that the increments contain nothing of their environing culture, past and present, except what is contained in the scientific text. Thus science is said to be different in kind from all other dimensions of culture.

My own view is that meaning in science is indeed less dependent on general cultural background and personal idiosyncracy than meaning in works of art or in religious beliefs. There is a quantitative difference in the extent to which interpretation must be added to texts of scientific work when these are contrasted with literary texts; but the difference is not qualitative.

Written communication eliminates the direct relationship between symbol and referent, a relationship which is the gist of oral communication. In oral discourse, there is no accumulation of successive layers of historically validated meanings. Instead, each word gains meaning in a succession of concrete situations, thus accumulating in the listener; this brings about the direct relationships of symbol and referent relationship. The positivistic view attributes to the written scientific text the properties of objective, unproblematic summary of accumulated knowledge. This contrast between oral and written communication leads to the wrong conclusion that scientific knowledge grows by sheer accumulation. The mistake lies in the fact that there are two qualitatively different kinds of accumulation: when an experimenter describes an experiment, the listener absorbs the story together with the "irrelevant" information about the personality of the speaker, the atmosphere of the meeting, the unimportant details of the specific experiment; nonetheless, he understands much of what is told even though the account is unclear, inaudible, or ungrammatical. If the experiment is written down in scientific language, all the previous irrelevancies are excluded. Yet there remains a major impediment to "total independence"; the text could not be written without the writer's "background knowledge" and it could not be read without the reader's "background knowledge." Creativity enters into what the individual makes of the written text.

It could in fact be said that written texts are more inclined towards obsolescence than oral "texts" because these are always open to reinterpretation. This may be one factor in the conservation of nonliterate cultures. Critical analysis is possible only of written texts. This is analogous to learning previously unexpected truths in a mathematical deduction from some physical premises. This is what the students of literacy call the autonomy of written statements. In prealphabetic writings, as in syllabaric systems, no novel statements can be formulated; they serve only to recall the already familiar.

The limits to the self-sufficiency of a scientific text and its dependence on interpretation may be illustrated by reference to a closely-knit scientific community, like the scientific and technical staff of a huge particle accelerator such as the SLAC or the Cambridge Linear Accelerator. In such a group, there are more shared symbols beyond the results which are codified in writing by the authors of a given paper. The shared beliefs, symbols, technical skills, the "images of knowledge" are not included in the publication. Differences in the interpretation of results resulting from divergent scientific metaphysics (espousal of different theories of elementary particles) come now to the forefront. In order for the differences not to impede the "progress of science" the text must contain only that on which all the authors agree. Much is left out; much must be added in interpretation.

The claim for the autonomous intelligibility of meaning in science (i.e. of the text of a scientific work), should be taken as an ideal to be striven for, rather than as a description. The objective in all science is to reach the stage where the "meaning is in the text." The consensus which every scientist seeks to gain around the propositions he puts forth might be more easily attainable. If our aim is the progress of knowledge (i.e. the advancing of new theories), or learning of new facts, or getting closer to reality, is the aspiration to this ideal fruitful? Can it lead to scientific "progress"? In the early stages of development of a theory or of a new discipline, when only a few individuals share the context for understanding "correctly"—in the new way—some of the theoretical or experimental findings, the approach to greater consensus is very important; it is conducive to the progress of knowledge. The scientists to whom the new ideas are addressed read and necessarily "misinterpret," according to their previous knowledge, what they read. The innovators respond by correcting, perhaps attuning themselves to some of the old concepts or views which they had not thought of when they first presented their new ideas. The readers of the "old school" read again and assimilate some of the new explanation; the result is a greater sharing of context. This goes on until at some point the belief spreads among the members of the given scientific community that a fruitful new theory has been established. It remains, however, ambiguous enough so that some persons working in the field can differently absorb the written text sufficiently for some original further contribution, for example, an idea or measurement or reinterpretation that is fruitful for progress which no one else has thought of. From this stage on, the common aim of arriving at a complete consensus by an agreed formalization and by reducing the possibility of interpretation as much as possible might lead to "dead text," at which point that branch of science ceases to grow. The ideal of a scientific text which is self-contained may never be realizable. The act that it is striven for, aspired to and that it is approximated does constitute a major feature of modern Western scientific thought in constrast with that of other non-scientific cultures.

The approach that accepts that the meaning is in the text relies exclusively on epistemic reason. Once we admit that this is not so, we can get at the multiplicity of meanings which are all in a sense 'in the text' by way of *cunning reason,* which then enables different individuals to find different meanings in the 'same' text and to make different discoveries. Progress is possible only due to cunning reason.

Second Order

Some eminent anthropologists see in the ability of Western culture to view itself reflectively the hallmark of scientific thinking, or second-order thinking. It is typical of such objectification and reflexivity to rely on strict logical discourse. On the other hand, a nonreflexive first-order approach to nature, society or the individual is typically interactionist: it applies cunning reason. It attempts to convince, cajole, manipulate nature or society or the individual. It seems to me wrong to see in second-order reasoning the chief characteristic of Western scientific thought. All justification is second-order and all discovery is first-order. There is no culture nor any discipline which has only second-order or only first-order thinking. Nor is it necessarily a sign of progress when second order predominates; it often signifies decadence or plain empty formalism occurring when an area of research stumbles on major difficulties.

First-order and second-order thinking, cunning reason and epistemic reason complement each other and occur parallel in every discipline and every culture.

REFERENCES

1. ELKANA, Y. 1978. Two-Tier Thinking: Philosophical Realism and Historical Relativism. *Social Studies of Science,* **8:** 309–326.
2. EPICURUS: "The Menoecus" in *Epicurus, the Extant Remains,* Oxford 1926, part 134. Quoted and adapted by John Passmore. 1978. *Science and its Critics:* 29. Rutgers University Press.
3. BENJAMIN, W. 1973. Understanding Brecht, *NLB:* 8.
4. LECKY, W. E. 1975. *History of European Morals from Augustus to Charlemagne,* Vol. 1. Arno Press, New York.
5. DETIENNE, M. & VERNANT, J. 1978. *Cunning Intelligence in Greek Culture and Society:* 3. Harvester Press, Sussex.
6. In anthropology a term developed by GEERTZ, C. 1973. *Interpretation of Cultures.* Basic Books.
7. REICHENBACH, H. 1936. *Experience and Prediction.* University of Chicago Press.
8. Even as late as 1967 Bar-Hillel finds the main difference between Carnap and Popper in their different emphases: Carnap on justification and Popper on discovery. This also

explains in Bar-Hillel's view the repeated misunderstandings between the two. He says:

"... let me first try to formulate clearly the difference of interest between Popper and Carnap in their treatment of science. In a nutshell it seems to be this: Popper is primarily interested in the growth of scientific knowledge. Carnap in its rational reconstruction; or, to borrow terms from current methodology of linguistics: Popper's philosophy of science is diachronically oriented, Carnap's is synchronic. With still other metaphors, Popper's conception is dynamic, Carnap's is static."

From: "Popper's Theory of Corroboration" *In: The Philosophy of Karl Popper.* Schilpp, Ed. Vol. **1:** 332–348. Open Court. 1974.

9.   "The Search for Philosophic Understanding of Scientific Theories," *In: The Structure of Scientific Theories.* F. Suppe, Ed.: 127. University of Illinois, 1974. Suppe's book, in addition to being a very good summary, is also very interesting in the attempt, typical of the generation, to bridge somehow the old positivism, and extracting the best from it, and the newer approaches.

After the above quotation follows a thorough analysis of the work of Hanson, Toulmin, Kuhn and others on who developed alternatives to the "Received View."

10. DOUGLAS, M. 1975. *Implicit Meanings.* Routledge & Kegan Paul, Boston.
11. ELKANA, Y. 1975. *The Discovery of the Conservation of Energy.* Harvard Univ. Press.
12. MEYERSON, E. 1962. *Identity and Reality.* Dover. Especially the chapters on 'Law and Cause' and on 'The Irrational.'
13. DOUGLAS, M. Ref. 10: xiv.
14. DOUGLAS, M. Ref. 10: xix.
15. BACON, F. 1964. Temporis Dartus Masculus. *In: The Philosophy of Francis Bacon,* B. Farrington, Ed.: 72. Liverpool University Press.

SCIENCE AND THE PROBLEM OF AUTHORITY IN DEMOCRACY*

Yaron Ezrahi

*The Hebrew University of Jerusalem
Jerusalem, Israel*

THE CRITICAL AND CONSTRUCTIVE FUNCTIONS OF SCIENCE

Since the rise of modern science in the 17th century, its social perceptions have been mediated by two related but distinct images. On the one hand, science has emerged as a criticism of sacred beliefs and myths, the negation of established traditional modes of knowledge and scholarship; on the other as the progenitor of a new tradition of knowledge and scholarship, the affirmation of certifiable beliefs and ideas. The one image projects the scientist as an iconoclast challenging established authority; the other projects him as a cooperative self-controlled individual scrupulous in adhering to the discipline imposed by the authoritative canons of the scientific method. The iconoclastic image of science brings to mind Galileo's confrontation with the authority of the Church or the tremors induced in the Christian world view by Darwin's evolutionary theory. The image of science as a restrained, moderate and constructive enterprise was conveyed in Bacon's vision of the scientific community in the *New Atlantis* and in Thomas Sprat's historical account of the early years of the Royal Society.[1]

Across diverse cultural and political contexts, the social images of science as a *critical* or a *constructive* enterprise have had different imports. At times the scientific criticism of established beliefs has been viewed as the assertive voice of freedom against oppression, at times as the manifestation of heretical revolt and potential disobedience. As the affirmative construction of a domain of knowledge, science has been identified at times with the noble objectives of enhancing tolerance and tempering the fanatic zeal of doctrinal conflicts, and at other times as but a form of communal terror which, under the guise of impersonal and objective truth, threatens to suffocate personal fantasy and undermine faith.

In support of each of these different perspectives on science, advocates and critics alike have cultivated selective notions of what the essential sociocultural features of science are which indicate its spirit, objectives and effects. For some, what made science suspect was the skepticism of a Descartes. For others it was the optimism of a Condorcet. For others the Baconian tradition in science and the postulates of observation and empirical verification were just the manifestation of science's role in the despiritualization of culture. For others the essence of science was indicated instead in mathematics which made science appear too remote from the common experience, inhabiting a domain of certainties irrelevant to the uncertainties of everyday life.

The history of the role of science as a cultural symbol in politics is largely a record of recurrent shifts between such different perspectives on the iconoclastic-critical and the constructive social images of science; between science's perceived role in debunking established beliefs and authorities and in evolving certified knowledge as a basis

*This paper is part of a larger work on *Science and Politics in the Modern Democratic State* supported by the Center for Advanced Study in the Behavioral Sciences, Stanford, The National Science Foundation (BNS-76-22943), and The Ford Foundation. The author is deeply indebted to their generous support of this study.

43

for the authoritative resolution of conflicts of opinion. In light of these apparently conflicting political "appropriations" of science as a symbol, it is not surprising that science could be both criticized by the Church in the name of established authority, as in the case of Galileo, and by radical and yet diverse spokesmen like Jean Jacques Rousseau and William Blake in the name of individual freedom.

It is perhaps a special feature of the liberal-democratic tradition that insofar as it regards order as grounded in individual freedom and is committed to a concept of society consistent with the sacredness of individual personality, its appropriation of science as a cultural symbol in politics encompasses both the critical and the constructive dimensions of science. For liberal-democratic thought the attraction of science as a cultural enterprise lies precisely in the special relations between criticism and certification; in the notion that criticism as a form of conflict is *the method* through which the authoritative rational consensus on truths is produced;[2] that in science, criticism, insofar as it serves the advancement of knowledge, is simultaneously a negation and a constructive affirmation of authority.

As a distinct universe of discourse and a particular form of association of men, science, then, has not been regarded only as the cooperative search for a special kind of knowledge. In the context of politics one can study the place and impact of science as an example or even a test demonstrating the feasibility of a set of cultural norms of discourse and association which are socially perceived as relevant for the ways people organize and conduct their public life.

Although the cultural features of science and liberal-democratic politics and their affinities have undergone profound transformations during the last three centuries or so, these formations continue in some important respects to be but developments or variations in the same cultural traditions. Underneath the differences and changes the continuity in affinities between the scientific and the liberal-democratic political traditions is manifest in a number of ways. In both traditions the rejection of personalism, transcendentalism and other forms of publicly unconditioned principles of authority has oriented the ambivalence towards authority; in both, a commitment to a method of discourse which leaves room for open criticism and a clash of opposing opinions sets the boundaries of the legitimate range of responses to the problem of restraint and control; and with respect to both, imbalances between the critical and the constructive functions of free discourse have appeared as endemic threats to the fusion of freedom and authority in the respective enterprises of science and politics.

This is not, of course, to imply that the cultural norms of discourse and association that characterize science and liberal-democratic politics are identical or perfectly harmonious. The advancement of knowledge and the construction and maintenance of public order are tasks sufficiently distinct as to generate quite different criteria for the adequacy of modes of discourse and association. In some respects, however, such differences have added to rather than diminished the power of science as a cultural symbol in politics. On the one hand, John Stuart Mill could observe in his essay *On Liberty* that both natural science and politics differ from mathematics insofar as the

> peculiarity of the evidence of mathematical truth is that all the argument is on one side. . . .
> There are no objections and no answers to objections.

This, observed Mill, separates mathematics from natural philosophy where "there is always some other explanation possible of the same facts—some geocentric theory instead of heliocentric, some phlogiston instead of oxygen."[3] But if the possibility of "objections and answers to objections" is what puts natural science in a class with politics, ethics and other such subjects as against mathematics, these subjects, continued Mill, are "infinitely more complicated."[4]

The point is that the difference in degrees of complexity not only detracts from the comparison but also adds a special dimension of relevance to the example of science in politics. As a simpler exemplification of the principles of discourse in which open criticism and diverse positions are resolved in authoritative consensus, science becomes an ideal for politics—an ideal toward which politics can only aspire. "In the free cooperation of independent scientists," observed a scientist-spokesman of liberal ideas a century after Mill,

> we shall find a highly simplified model of a free society, which presents in isolation certain basic features of it that are more difficult to identify within the comprehensive functions of a national body.[5]

In the early 20th century, the belief in the centrality of the example of science and its didactic power in upholding and spreading liberal-democratic principles is, of course, prominent in the thought of John Dewey. Dewey observed that "the operation of cooperative intelligence as displayed in science is a working model of the union of freedom and authority" relevant to the political, economic and moral spheres.[6] He insisted that the very

> future of democracy is allied with the spread of the scientific attitude . . . it is the only assurance of the possibility of a public opinion intelligent enough to meet present social problems.

It is also the only guarantee, observed Dewey, against "wholesale misleading by propaganda."[7] Against the mounting assaults launched by anti-democratic political movements and ideologies of the period between the two World Wars, Dewey seeks to reinforce liberal democracy by extending to the larger social context the "scientific moral" code which he identifies with a specific "method of discussion" or procedure of organized cooperative inquiry that has led to the "triumphs of science in the field of physical nature."[8]

The force of science as a simplified and clearer exemplification of the ways liberal-democratic principles of discourse work in their purest form has also been a liability. The symbolic and didactic functions of the example of science in liberal-democratic politics have opened the way to making science a convenient target for the critics and opponents of liberal-democratic politics. It has also meant that any gaps between the socially visible practice of science and its ethos, any socially perceptible failure of science to exemplify liberal-democratic principles of discourse and association were bound to take on a cultural and political significance which extends far beyond the status of science as a social institution. Where scientists are perceived not as the politically neutral custodians of universal intellectual norms but as partisan advocates whose claims of objectivity conflict with their apparent political and institutional affiliations; when the methods devised to separate subjective personal opinions from truth and distinguish facts from beliefs are socially perceived to be ineffective even in the domain of scientific discourse, then it is not only the social authority of science which is undermined. Perhaps no less consequential are the corrosive effects on the faith in the very feasibility of a mode of discourse which in harmonizing individual freedom, reciprocal criticism and rational consensus was identified with the ideal liberal-democratic solution to the problem of moralizing restraint in a free polity.

Before turning to a more detailed discussion of the socio-cultural bases of the symbolic role of science in liberal-democratic politics, I should like to anticipate a possible objection to the preceding comments. The failure of science to socially and culturally confirm the validity and feasibility of liberal-democratic principles of discourse and association need not require that the socially perceived discrepancies between the practice and ethos of science be "real" any more than the symbolic power of science in liberal-democratic politics requires that such discrepancies must in fact be absent. The symbolic functions of science depend upon both the practice and the socio-cultural perceptions of science. At times, the social perceptions may be sufficiently independent of the facts of scientific practice to warrant the observation that the fluctuations in the cultural role of science in politics and society may be more successfully traced to changes in the socio-cultural environment of science than to its internal organization. Nevertheless, with respect to the declining symbolic power of science in supporting liberal-democratic principles, especially since World War II, developments in the organization of science and its socio-cultural environments seem to have converged in rendering particularly damaging what appeared at first to be only another in a series of periodic attacks on the validity and the generalizability of the ethos of science. In order to appreciate the political significance of this most recent upsurge of anti-science sentiment, it is necessary first to examine more closely the ethos of science and its affinities to liberal-democratic notions of discourse and association.

THE ETHOS OF SCIENCE AND LIBERAL-DEMOCRATIC IMAGES OF KNOWLEDGE AND AUTHORITY

In a period in which the belief in the eternal validity of truth has been shaken by the discovery of the historicity of scientific thought, it may be striking to reveal the remarkable thematic continuity in the ethos of science. This continuity becomes instantly apparent when we juxtapose the 17th century exposition of the norms of the Royal Society by its first historian, Thomas Sprat (1667), and the 20th century discussion of the ethos of science by the pioneering American sociologist of science, Robert K. Merton (1938, 1942). In 1942, at a time of upheaval and political, ideological and military conflict, Merton published an article entitled "Science and Technology in a Democratic Order,"[9] in which he laid down the basic elements of the ethos of science. The background of political instability adds a particular dimension to the comparison with Sprat, who in 1667 published his *History of the Royal Society* against a similar background of social conflict and political upheaval. Although in 17th century England political instability was largely induced by religious disputes, while the Western democracies of the 1930s and 1940s were beset by ideological conflicts, both these historical settings have rendered the ethos of science politically relevant in similar ways.

In his 1942 essay as well as in an earlier complementary paper entitled "Science and the Social Order,"[10] Merton, to be sure, does not specifically address himself to the political uses of the ethos of science or to the political implications of the social criticism of science. Merton touches on these issues, but his concern is primarily with the kinds of social and institutional contexts which are congenial to or restrictive of the autonomy and the free development of science. Yet although, unlike Sprat, Merton did not use a discussion of the normative structure of science to recommend its generalization into other cultural spheres, such an extension was made by Merton's contemporaries, the most prominent of whom was John Dewey. In both historical contexts, then, discussions of the characteristics of science interpenetrated debates concerning the desirable norms of discourse and association in the civil society.

Merton has identified four institutional imperatives in the scientific enterprise: universalism, communism, disinterestedness and organized skepticism. Universalism finds immediate expression in the canon that truth-claims, whatever their source, are to be subjected to pre-established impersonal criteria consonant with observation and previously confirmed knowledge. . . . The acceptance or rejection of claims in science is not dependent upon any personal or social attributes of the participants. The imperative of universalism is deeply rooted in the impersonal character of science and in the nature of science as a cooperative venture.[11] In Sprat's description the Royal Society has adopted similar rules. The enterprise is not to be trusted to single men . . . "not to philosophers; not to devout, and religious men alone."[12] In the Royal Society

> . . . the Soldier, the Tradesman, the Merchant, the Scholar, the Gentleman, the Courtier, the Divine, the Presbyterian, the Papist, the Independent and those of Orthodox Judgment, have laid aside their names of distinction and calmly conspired in a mutual agreement of labors and desires . . .

This, observed Sprat, is a blessing "which seems even to have exceeded that evangelical promise that the Lion and the Lamb shall lie down together . . . ."[13]

Merton's second norm of science is communism, which signifies the attitude that regards the "substantive findings of science" as a "product of social collaboration" and a "common heritage."[14] Merton relates the imperative of sharing the fruits of research to the imperative of communication of findings and the rejection of the norm of secrecy.[15] In Sprat's description, the scientific enterprise exemplified in the practice of the Royal Society a collaboration based on a "union of eyes and hands."[16] Anticipating such modern cultural institutions as public museums and public libraries, he even saw it as one of the objectives of the Royal Society

> to purchase . . . extraordinary inventions, which are now close locked up in cabinets; and then to bring them into one common stock; and which shall be upon all occasions exposed to all men's use.[17]

Merton's third norm is disinterestedness, which is linked with the "public testable character of science"[18] and is exemplified in the various ways which regulate the "accountability of scientists to their compeers."[19] In Sprat's description it is a "plain industrious Man" who is "more likely to make a good philosopher than all the high earnest insulting wits who can *neither bear partnership nor opposition.*"[20] (My emphasis). The Fellows of the Royal Society sought, by his account, to free the knowledge of nature from "humors and passions of sects."[21] Because

> men of various studies are introduced . . . there will be always many sincere witnesses standing by, whom self love will not persuade to report falsely, nor heat of invention carry to swallow a deceit too soon; as having themselves no hand in the making of the experiment but only in the inspection.[22]

The norms of universalism, communism and disinterestedness in Merton's account and their near equivalents in Sprat's can be seen as elements in what I have called the *constructive* function of science. They represent different aspects of science as a cooperative enterprise in which free individuals voluntarily socialize their work. Merton's fourth element of the scientific ethos—organized skepticism—and its counterpart in Sprat belong mainly to the *critical* function. Merton links "organized skepticism" as "a methodological and an institutional mandate" with conflicts between science and other institutions and eruptions of *social* resentments towards science.[23] Sprat, who is so effusive in describing the constructive function of the scientific activity, is more sparing and cautious in relating to its critical function. It is, however, unmistakably there in his distinction between "the knowledge of Nature" and "the colours of Rhetoric, the devices of fancy or the delightful deceit of Fables."[24] It is clearly manifest in his insistence that "the seats of knowledge" have been for the most part "not laboratories as they ought to be; but only schools where some have taught, and all the rest subscribed."[25] It is also detectable in his defensiveness against the imputation of atheism to the members of the Royal Society[26] and in his comparing of the Royal Society to "our Church in its beginning." "Both," said Sprat,

> may lay equal claim to the word Reformation . . . they both have taken a like cure . . . passing by the corrupt copies and referring themselves to the perfect originals for their instruction: the one to the Scripture, the other to the large Volume of the Creatures".[27]

If skepticism directed against outside systems of belief and authority expresses the critical function of science, when it is viewed as an imperative of the interaction among scientists it is part and parcel of the constructive function of producing the consensus which certifies scientific knowledge. It is this constructive dimension of criticism in the internal life of science which lends special force to science as a cultural example relevant to the liberal-democratic response to the problem of authority.

The constructive function of criticism in science has a variety of manifestations. One of the principal ones involves the elevation of the eye in the "communication of light."[28] What Sprat calls the "Union of eyes" is first of all a means for the criticism of "doctrines without a sufficient respect to works"[29] and for restraining the effects of the "glorious pomp of words."[30] But the rejection of the ornaments of speech by "solid practice and examples"[31] rests also on the constructive role of the eye as a medium through which the new men of knowledge are "very positive and affirmative in their works."[32] In this early formulation of the "cultural code" of science, the most revolutionary development was the discovery of the eye as a binding social force,[33] the "redemption" of vision from the domain of relative subjective experiences* and its cultural redefinition as an objective, positive component in the cultural construction of the public realm.

Even prior to Sprat the move to socialize the eye as a primary medium of impersonal public communication was manifest in Bacon's preference for the "Initiative" over what he called the "Magistral" method of transmission. Whereas the "magistral" method requires, according to Bacon, that "what is told should be believed, the initiative [requires] that it should be examined."[34] Bacon scholars have pointed out that in Bacon the force of the inductive method, the superiority of examining over telling, is not only in the advancement of knowledge but also in being a stronger and more persuasive mode of presentation. "All method is for Bacon to a greater or lesser extent an artifice for convincing presentation."[35] Reliance on trials and "Works" rather than on "words" is not only a better way to the truth; it is also the more convincing "rhetoric."[36]

Juxtaposed with Bacon-Sprat's emphasis on the "union of eyes" there is, of course, the other scientific tradition for which the binding force of knowledge lies in its mathematical skeleton even more than in its foundation in common observations. If for Bacon and Sprat what makes natural philosophy binding is the shared examination of works "as pledges of truth,"[37] for Galileo it is the certainty of necessary mathematical demonstrations which (even more than the testimony of the senses) distinguishes science from all other fields. "Every expression of Scripture is not tied to strict conditions like every effect of nature" observed Galileo in his letter to the Grand Duchess of Tuscany. He called the "prudent Fathers" to consider ". . . the difference that exists between demonstrative knowledge and knowledge where opinion is possible (and) . . . with what force necessary conclusions compel acceptance . . . It is not in the power of those who profess the demonstrative sciences to change their opinions at pleasure and to apply themselves now on one side and now on the other." For Galileo

---

*"Every Eye sees differently. As the Eye such the object," wrote William Blake (from Annotations to Sir Joshua Reynolds' Discourses, London, 1798).

the compelling authority of necessary demonstration was, of course, a shield against the intrusions of censorship. "The demonstrated conclusions touching the things of heaven cannot be changed with the same facility as opinions about what is legal or not . . . ."[38] Ironically, the same rationale employed by Galileo to restrain the intervention of the authority of the Church in scientific matters would be employed in later years to protect science and academic freedom against the intrusion of external authorities including the democratic authority of free public opinion.

But if the scientific revolution was largely the result of a fusion of powers of mathematics and empirical observation at least in the Anglo-American context, the liberal-democratic appropriation of science as a symbol of the ideal culture of politics focused selectively upon the latter. As is indicated in John Stuart Mill's classification of both natural philosophy and politics apart from mathematics,[39] liberal-democratic visions of politics were not inspired by the nondiscursive Galilean notions of the authority of mathematical certainty and necessary conclusions but rather by empirical tradition; not by the relevance of the authority of truths which are demonstrable without debate in the court of individual reason but by the power of visual persuasion controlled through communal testimony and further certified and socialized through discourse.

At the deliberate risk of simplifying a complex matter, I would like to dwell a little further on the politically relevant differences between the characteristics of mathematical and empirical persuasion. As two distinct modes of validating claims, mathematics seems to rest truth on *inferences,* whereas empirical truth rests on *testimonies.* Whereas the certification of mathematical proofs does not increase significantly with the accumulation of repeated attempts to retrace the steps leading to the result, with reference to empirical proofs the accumulation of testimonies is more significant for furthering certification and adding weight. If with respect to mathematical certainty the utility of replication lies in the usually remote chance of discovering an error, with respect to empirical truths the utility of the accumulation of testimonies is in affecting the distribution of probabilities among competing statements. Even with repect to empirical truths there is, to be sure, a point beyond which the endless repetition of the same experiment may be as futile as repeating the same proof in mathematics. Moreover, when compared with traditional concepts of knowledge based on revelation, special access to secret texts or languages and other forms of exclusive knowledge, both mathematical and empirical concepts of knowledge decentralize or democratize the community of rightful and potential knowers. *But whereas mathematics tends to induce a sort of radical decentralization that minimizes the need for the individual man of knowledge to rely upon and collaborate with his peers, empirical knowledge both decentralizes and communalizes.* This observation concurs with Mill's point that in mathematics, where difference of opinion is less acute, "all the argument is on one side," whereas natural philosphy, like liberal-democratic politics insofar as it progresses through the juxtaposition and the comparison of alternative positions, is more fully discursive.[40]

The evolution of scientific constructions of reality from a multiplicity of mutually corroborative and corrective individual testimonies and the certification of empirical truths through impersonal technical discourse, teamwork and comparisons, have made the empirical tradition in science a cultural model particularly relevant to the task of balancing the individual and the public realms in liberal-democratic politics. By comparison with the relatively lower significance of debate and collaboration in mathematics, the experimental sciences appear as a more perfect example of a cultural enterprise in which the construction of communally binding, intersubjective standards rests upon the coordination of individual contributions. It is also, therefore, a more relevant example for the problem of fusing the critical and the constructive functions in liberal-democratic politics. This preference for the empirical-experimental elements in the scientific tradition is, of course, further reinforced in liberal-democratic thought by its long-standing suspicion of the Platonic use of a mathematical concept of knowledge to support a hierarchical principle of political authority. The historical analogy between mathematics and Latin as technically demanding and esoteric languages of exclusive communities of scholars could only increase the antipathy.[41]

The idea of observable reality as both the consequence and the epistemological matrix of a *public meeting of a multiplicity of individual minds* was clearly more akin to the liberal-democratic commitment to the idea that authoritative knowledge must be shared, and hence public, knowledge, that only the claims of knowledge which can be confirmed by reference to intersubjective experience can oblige. There is perhaps no more apt formulation of the nature of the social sharing of knowledge of facts than the one given by Hobbes:

> When two, or more men, know of one and the same fact, they are said to be conscious of it one to another, which is as much as to *know it together*. And because such are fittest witnesses of the fact of one another, or of a third, it was and ever will be reported a very evil act, for any man to speak against his conscience: or to corrupt or force another to do so. [My emphasis][42]

Hobbes' political theory, admitting a central place to the "conflict of interests" model of politics, was naturally driven to skepticism with respect to the power of both mathematical and empirical truths in restraining conflicting political opinions. Even the truth of geometry, he said, will be sustained only as long as it "crosses no man's ambition, profit and lust."[43] Nevertheless, in his description of the communal dimensions of empirical knowledge as a form of "knowing together" and his observation that the shared knowledge of facts moralizes the commitment not to deny these facts, Hobbes touches the heart of the symbolic link between science and liberal-democratic politics.

It is just such a "knowing together" which was projected in Sprat's description of the Royal Society and its Fellows as an expanding community of "witnesses of facts." If mathematical or geometric knowledge suggested the possibility of a multiplicity of discrete individuals *"knowing separately the same truths,"* empirical knowledge as a form of *"knowing together"* the truth was a closer approximation of the working of the liberal-democratic community. This is not to suggest, of course, that Sprat was a liberal-democratic spokesman, but rather that in his attempt to defend the Royal Society in the socio-cultural context of his time, he projected science as a cultural enterprise showing great affinity to what has come to be recognized as liberal-democratic principles of discourse and association.

Insofar as the self-restraint exercised by scientists who know the truth together was not regarded as an automatic external check imposed upon self-interested individuals by an "invisible hand," science was in very important respects superior as a cultural model for liberal-democratic politics to the free market mechanism in classical economic thought. Unlike the economic man, whose rationalized egoism ensures in the economic model the harmony of individual interests, the scientist is a public-regarding individual with an explicit moral commitment to public values. Merton's emphases on the communal sharing of the fruits of individual contributions in science (the norm of communism) and the impersonal, publicly testable character of science (related to the norm of disinterestedness) are pertinent here.[44] These norms accentuate the communal public values which distinguish scientific from economic activities and render the former a more compelling cultural model for the constructive task of liberal-democratic politics. These cultural differences are perhaps detectable also in the differences which are characteristic of the metamorphoses of scientific and economic achievements into respective assets of political leadership. Whereas business leaders who seek public office or influence are often led to make demonstrable contributions to public values to counter their presumed identification with the values of utilitarian egoism, scientists as public opinion leaders, at least until they have been tainted by partisan institutional affiliations or by the negative consequences of their discoveries, are already identified with public values in their initial professional posture as scientists.

The empirical image of the advancement of knowledge as a public enterprise of free individuals could then appear superior to both the cultural models of mathematical knowledge and the market mechanism in instructing liberal-democractic orientations towards the construction of authority. It appears to ground authoritative standards in a system of decentralized individual participants, fostering self-restraint through voluntary communal accountability. Science could thus come to repesent the feasibility of restraint less blemished than in other spheres of activity by transcendental, factional, personal or other publicly inaccessible references.

The apparent characteristics of science as a form of *irenic* rather than *polemic* discourse; the image of scientific discourse as an example of the tolerant and moderate use of language free from the zeal and enthusiasm of unresolvable doctrinal disputations or personal clashes which easily degenerate into violence, could only reinforce the positive symbolic import of science as a model of peaceful resolution of conflicts. Against a background of violent verbal confrontations, the "silent, effectual and unanswerable arguments of real productions," the reliance on experiments and observations in the resolution of disagreements, were bound to have a special attraction.[45] The scientific mode of discourse was to Sprat one remedy to the strife induced by political and "spiritual" controversies. It could contribute, in his opinion, to "the sweetening of such dissentions."[46]

The discipline displayed in scientific discourse could thus appear to confirm one of the basic epistemological presuppositions of liberal-democratic politics: the feasibility of extracting from subjective worlds of perception and experience the objective communicable elements which could furnish the building blocks of an intersubjective, objective sphere. Against the often shaky alternative of resting the public realm on the premise of an ultimate harmony of values or interests, science could appear to represent the possibility of supporting the constructive enterprise of liberal-democratic politics by homogeneous epistemological standards. Insofar as the organizational principles of the scientific enterprise sanctioned the voluntary standardization and socialization of individual observations and grounded empirical statements in the concurrence of a multitude of mutually accountable individual testimonies, the social authority of science could appear to confirm the liberal-democratic method of moralizing restraint in a manner compatible with freedom. By virtue of its characteristic as a sort of "public knowledge," science could evolve "truth" and "reality" not only as the fruits of a collaborative intellectual enterprise, but as categories of legitimate limits, limits on human conduct which rest neither on arbitrary personal judgments or transcendental references, nor upon the form of dogmatism which, according to Sprat, produces "a company all of one mind."[47] As forms of moralized restraint, scientific truth and scientific constructions of reality appear to be conditioned by the public, that is, by concurring, mutually accountable but independent individual testimonies. The authority of science could therefore be regarded as the result of a decentralized cooperative enterprise which discovered the methods through which freedom breeds discipline.

This view, that as a socio-cultural enterprise science embodies liberal-democratic values, tended to find confirmation in the nonhierarchical collegial and international structural properties of the scientific community.[48] Consistent with these values, liberal-democratic ideals of individual autonomy appeared to be manifest also in the context of scientific education. Experiments, observed Sprat, are substituted for notions so that in them even children "may soon become their own teachers."[49] The students of scientific subjects, just like the citizens of liberal democracy, abide only by authority which they themselves have constituted. As self-constituted, legal standards or empirical truths presuppose authority which rests on autonomy, not dependence.

The analogy to social contract theories is, of course, very suggestive. But I would like to emphasize in this discussion its relevance not with regard to the genesis of public civic culture but with respect to the symbolic import of science in validating the epistemological presupposition of such a culture. The socially demonstrable success of science, which assumed visible technological dimensions especially in the fields of physics, chemistry and medical biology, and its unquestioned validity across cultural and social boundaries, have appeared to confirm one of the most cherished presuppositions of liberal-democratic politics: that the cognitive-perceptual equipment of man is universally uniform. The history of the belief in the universal uniformity of human cognitive and perceptual equipment and its many expressions in liberal-democratic political thought and ideology is, of course, both rich and complex. But one need not dwell on the matter in order to recognize the architectonic function of this belief in the political enterprise of the liberal-democratic tradition, particularly with respect to the goal of constructing the public realm on the multiplicity of *commensurable individual judgments*. In relation to this task, the socially and universally demonstrable validity of scientific knowledge has often been assimilated in liberal-democratic arguments as ongoing confirmation of the reality of an intersubjective public sphere of cognition and of the presence of an epistemological domain which could constitute the matrix of socially binding standards. Taken as a testimony to the universal uniformity of the cognitive-perceptual capacity of man across the diversity of environments, cultures and values, science has furnished essential support for the liberal-democratic commitment to anchor political order in cooperative individualism rather than in organizationally centralized power.

The association of the "culture of science" with images of the modern liberal democratic society is, of course, pervasive in modern sociological and ideological discussions about actual and desirable models of man and society. One instance is the work of the influential social thinker Ferdinand Tönnies at the end of the last century.[50] Tönnies distinguished between two types of human association—"community" and "society" (*Gemeinschaft and Gesellschaft*). In his analysis, community is a human association which revolves around such social foci as the family, the village and the town, and its cultural fabric consists of folkways, customs and religion. By contrast, society is a social organization whose characteristic basis is urban individualism and whose cultural mode is manifest in legal and scientific standards and cosmopolitan public opinion. The normative characteristics of the Gesellschaft (society) in Tönnies' description bear close affinity to what Merton has identified as the principal features of the ethos of science. Among the latter, Gesellschaft is guided primarily by the norms of universalism and disinterestedness. It is a form of human association in which social relations are regulated by impersonal-universalistic standards. The influence of such Tönniesian-like juxtapositions between community and society as distinct modes of social association is echoed in modern debates on the merits or demerits of the modernization of traditional societies. Considering the perceived affinity between modern urbanized societies and the culture of science, it is not surprising that the criticisms of the "culture of science" are so often manifestations of opposition to the values of social associations in which principal institutional principles of science are normative. Consistent with this view, even in totalitarian states (not exactly the traditional society to which Tönnies referred), the "ethic of science" is, as Merton observed, typically rejected and criticized as "liberalistic" or "cosmopolitan" or reflecting "bourgeois prejudices."[51]

The special affinity between the ethos of science and liberal-democratic images of society and the polity is indicated not only in the criticism of the culture of science in totalitarian and traditional societies or in the affirmation of science in the liberal-democratic tradition. It is manifest also in the kinds of criticisms directed against science from within liberal-democratic societies. Insofar as the liberal-democratic polity rests on a continual attempt to maintain a precarious balance between self and society, science, as a central symbol of public culture, inevitably gets entangled in the continual border clashes between the domains of the public and the self. Thus whereas in the totalitarian society the rejection of science is often only an act of political and ideological self-defense of the ruling elite against the force of universalistic impersonal standards of criticism, in the liberal democracy criticism against science is often an expression of the recurrent revolts launched by the individual against the strictures and constraints imposed on the freedom of the self in the name of public culture. In totalitarian states and in traditional societies, science tends to be resented primarily by virtue of its perceived challenge to the established principles of integration and order and the destructiveness of its critical function. In liberal democratic states, it is the perceived 'unbalanced' extension of the constructive function of science in the building of the public domain which is often resisted as an illegitimate intrusion into the domain of the self.

THE RELATIVIZATION OF SCIENCE AND THE CRITICISM OF PUBLIC CULTURE

In both totalitarian and liberal-democractic political contexts, strategies for eroding the social authority of science and relativizing scientific claims of knowledge and truth share, to be sure, similar traits. In both, the attacks have exploited the perceptible discrepancies between the ethos and the practice of science and in both, attempts to discredit the universalistic-disinterested social image of the scientific community have been carried out by questioning, *ad hominem,* the irrelevance of particularistic-ascriptive traits like race, faith, class or personality in the work of science. But within these generally shared lines of socio-political criticism, the difference between attacks launched on science in liberal-democratic and totalitarian states, respectively, lies in the type of particularistic traits ascribed to science in order to debunk the claims of objectivity, neutrality and rationality. Whereas in totalitarian political contexts like the fascist and communist regimes, scientists are suspect primarily as the representatives of illegitimate or discarded groups ("Jewish science", "bourgeois science"), in the liberal democracy where group diversity and conflict are more legitimate, a particular accent has been put on relativizing science by references to the weight of subjective psychological factors. The one has tended to undermine science as a class culture, the other as inseparable from the subjective worlds of individuals.

In totalitarian—or more accurately, doctrinaire—regimes, the social authority of scientists is most typically challenged by "exposing" suspect particularistic group affiliations as a way to discredit the claim of *universalism.* In liberal-democratic regimes, their authority is typically challenged by using allusions to suspect personal needs and motives as a way of casting doubt on the adherence to the requirement of *disinterestedness.* The point is not that we do not find both strategies of criticism employed in doctrinaire and liberal-democratic regimes. It is rather that the relative potency and the political appeal of these strategies vary across different political contexts according to the differences in the elements of the ethos of science which are central in the social legitimation of the authority of science. In totalitarian-doctrinaire regimes where science conflicts with the official doctrine as a competing code of public culture, the criticism of science cannot be based on the revolt of the self against the very idea of public culture. By contrast, in the liberal-democratic political systems, the "exposure" of the subjective psychological dimension of science is the typical strategy for the politicization of the authority of science, a criticism of the very claims of public culture in the name of the cherished values of personal freedom and creativity.

Merton made the apt observation that in "a liberal society, integration derives primarily from the body of cultural norms toward which human activity is oriented ... In a dictatorial structure, integration is effected primarily by formal organization and centralization of social control."[52] Merton further observed that such differences in the mechanisms through which "integration" is typically effected permit "a greater latitude for self determination and autonomy in various institutions, including science, in liberal than in totalitarian structure."[53] But if the very ethos of science is, as I have proposed, such a central part of the "body of cultural norms" which orient behavior in the liberal society, these externally added functions render the autonomy of science vulnerable in at least two principal ways. First, the external political support for the autonomy of science can suffer from internal changes in the relative position of the ethos of science in the established body of liberal-cultural norms. Second, the external ideological and political load of the ethos of science, insofar as it is regarded as a part of the normative code of the liberal-democratic polity, can render science a convenient target for critics who expound alternative cultural images of politics and authority.

In light of these possibilities, it is not surprising that, especially since World War II, debates concerning the objectivity, universality or rationality of science have so often interpenetrated the ideological polemic on the principles and structures of liberal-democratic politics. Consider, for example, the ideological ramifications of philosophical criticisms directed against classical positivist concepts of knowledge and rationality,[54] or the political import of the tendency of contemporary historians and sociologists of science to supplement "internal" history of science as the intellectual genealogy of ideas by an account which assigns greater weight to the "context" of scientific inquiry. The point is not that such philosophers or historians of science are politically biased. Even among those who clearly are, there are some who have made important intellectual contributions to our understanding of science. The point is, rather, that such works can easily lend support to the attempts to socially relativize and erode the claims that science is an enterprise directed by universalistic, impersonal and objective norms. The social and historical studies of science can be and apparently have been effectively used in the effort to debunk science as a symbol of the validity of public cultural standards and to upset the expectation that culture could constitute the principal balancing wheel between the countervailing demands of the domains of the public and the self. No wonder then that the exponents of new norms of liberal-democratic politics in which the boundaries of public culture and public authority recede before the assertive freedom of individuals and groups have typically added the works of people such as Feyerabend[55] and Thomas Kuhn[56] to their arsenal.[57] The extraordinary popularity of Kuhn and the wide ideological appropriation of the concept of "paradigms" to historicize and sociologize the social perception of the canons of scientific rationality—uses and misuses, against which Kuhn himself has often protested—reflect these developments. Further support for the evolving climate of skepticism towards the objectivity, disinterestedness and rationality of science has come from the works of people such as Daniel S. Greenberg[58] and James D. Watson[59] which are widely constructed as effective "exposés" of the "real" selfish personal motives which guide the conduct of scientists not only in the corridors of power, but also in the insulated temples of basic research. Especially in the 1960s, the cumulative effects of such books was to boost the suspicions that the practice of science is far removed from its ethos. The discrepancy appeared to suggest to some that "the moral values to which science formerly appealed are losing their sanction."[60]

The different lines of criticism which attack science as a way of undermining the cultural-epistemological foundation of liberal-democratic politics were consolidated by Theodore Roszak in one of the most articulate manifestos of the "counter culture."[61] Moving freely between the debunking of science and the challenging of the political orientations which assimilated its ethos, Roszak links skepticism towards the claims of "objectivity" with a celebration of the revolt of personality against authority. In the polemic concerning the "two cultures," Roszak typically poses himself as an apostle of humanism and literature against such voices as that of C.P. Snow, "the scientific propagandist."[62] In the spirit of Blake's "Vision of the Last Judgment", many of the rebels of the 1960s seem to experience the reality of "the outward Creation" as "hindrance and not action."[63] But whereas for Blake reality is transcended by consummating religious visions, the radicalism of the 1960s typically—though not exclusively—anchors the denial of constraints imposed by public and objectively certified experience in the subjectivity of psychedelic trips. In both cases, however, the individualization of the perception and experience of 'reality' has been canonized as a tenet in the epistemology of liberation from 'official public culture'.

At another level, the relativization of the social authority of science and the declining faith in universalistic public cultural standards were accelerated especially after World War II by the proliferation of open and continual scientific controversies.[64] The fact that these controversies tended to revolve more around questions of the use of science in public policy than around internal scientific issues has not mitigated the social effects of the discord in the ranks of the "new priesthood." The political relevance of these confrontations has only contributed to the visible signs of disharmony, and the sight of dissenting scientists in the glaring light of the public arena could only discredit the claim that the scientific community is an association of men who behave as "Gods one to another and not wolves."[65] Although the controversies have accompanied the expanding relevance and application of science to social and policy problems, they have appeared to many as merely the behavioral manifestations of the absence of universally objective and binding standards—a lacuna already noted by some philosophers, historians and sociologists of science.[66]

CONCLUSION: THE CHANGES IN THE LIBERAL-DEMOCRATIC CONCEPT OF POLITICS AND THE DECLINING RELEVANCE OF SCIENCE

If the observations concerning the place of science in the normative code of liberal democratic politics are valid, the mounting skepticism towards the canons of scientific rationality and the ethos of science can be regarded in many ways as indicating a significant political change. Inasmuch as the relativization of science is directed at denying the public nature of culture, it is also a weapon directed against the liberal democratic concept of culture as a mechanism for the authoritative integration and orientation of the civil society. Once culture, like religion in former times, is removed from the domain of the public to the domain of the self, where cultural norms do not represent shared communal but primarily subjective perceptions and judgments, culture is no longer a safe bridge between individual freedom and social discipline. When culture does not appear to civilize natural men by transforming them into citizens, it loses its power as a remedy for the problem of discipline in a free society.

From a historical perspective, the attack on science as an instrument for imposing illegitimate strictures on individual conduct and fantasy is ironic. Despite the fact that since the rise of modern science there have always been individuals and groups who warned against its cultural influences in the name of freedom, poetry, creativity and what not, the concerted contemporary attempt to debunk science is historically unprecedented. It is ironic if one compares the contemporary enthusiasm for disestablishing science by *ad hominem* references to the psychology, economics and politics of scientists, the sociology of the scientific community and the historicity of scientific truths with 17th—and 18th—century uses of science to inspire and direct the emancipation of the individual from the shackles of traditional authorities; or if one juxtaposes recent attempts to assert, against the norms of scientific culture, the counter-cultural principles of personality, authenticity and standardless creativity with the more distant early uses of science to free men from the authority of the past, the authority of the transcendental and the authority of dogma.

I have tried to point out that the historical alliance between the empirical, experimental tradition in science and anti-authoritarianism in politics was not confined to the context of criticizing and rejecting traditional establishments, that science, by demonstrating the productive capacity of the "union of eyes" and the "communal witnessing of facts" was integrated also as a model for the constructive task of building an alternative liberal-democratic political order. John Locke noted that "the decisions of churchmen whose differences and disputes are sufficiently known cannot be any sounder or safer than [the King's]."[67] But, stripped of the external controls of the Church, what was the alternative to the arbitrary will of the King? This was the problem confronted by modern liberal-democratic political theorists like Locke who sought a principle which could free authority from the bonds of transcendence without leaving it prey to unaccountable aribtrary wills. Between the publicly unconditioned despotism of a religious hierarchy and the tyranny of arbitrary personal wills, *science appeared to suggest a cultural strategy for depersonalizing authority through the free operation of a voluntary, self-regulating community which evolves universally valid standards.* The persistent normative force of science in the liberal-democratic tradition has stemmed precisely from the fact that the relevance of the ethos of science was not exhausted in supporting the liberal-democratic criticism of publicly unaccountable authority. Instead, science became also a cultural model for the positive function of liberal-democratic politics: constructing an alternative political order which rests on a balanced symbiosis between the public and the self, and the moralization of restraint by anchoring discipline in freedom and culture.

This constructive role of science in the liberal-democractic enterprise has now come under heavy attack. Twentieth-century controversies among scientists, especially in matters of public policy, are having effects on the discretion of government similar to the factionalisms and disputes of the churchmen in previous centuries. Compounded by the recent eruptions of passionate student radicalism on the campuses of Western universities, the conflicts of scientific opinion have undermined the faith in the power of science and the ethos of science to foster discipline. The discretion of political authority which since the 17th century has increasingly been freed from the external check of religious standards is again left increasingly undisciplined by the external control of authoritative standards of cognitive rationality. No longer confined by the epistemological presuppositions which enclosed liberal-democratic politics in a pendulum movement between criticism and consensus, the epistemological terrain from which objective public standards of truth have largely withdrawn does not give rise to consensus but, at best, to compromise.

This shift casts light on the contemporary crisis of authority in the liberal-democratic state as a loss of faith in public culture, the disintegration of the balancing wheel between the politics of criticism and the politics of construction. The decline of science as a cultural symbol of the public realm corresponds to the decline of public culture as a principal instrument for the constructive function of liberal-democratic politics. The trends eroding the cultural force of science in validating the epistemological and normative matrix of a public-intersubjective political enterprise threaten to leave criticism unbalanced by construction.

These trends, however, do not necessarily lead to chaos. In a polity in which the moralization of restraint is not effected by *truth* or by certified knowledge of reality, it may still be preserved by *representation.* In such a society the relation between science and politics undergoes profound transformation. In a society where politics does not aspire to evolve consensus but is satisfied with the more limited achievement of compromise, there cannot be much reverence for science as a cultural model for building consensus on truth. Ironically, the public, impersonal traits which formerly rendered science authoritative are precisely what make them suspect in such a polity. What formerly depoliticized science as a public enterprise now politicizes science as the antagonist of individual interests. The contemporary politicalization of the *social authority* of science is therefore one of the principal manifestations of the retreat of the politics of consensus before the politics of compromise. Whether the universalistic public authority of scientific standards is denied or relativized in the name of individual freedom and individual creativity which is extended beyond the realm of values to the cognitive construction of experience, or is undermined by allusions to the weight of the particularistic socio-psychological "contexts" of scientific inquiry, the results are the same. Public cultural values lose much of their authority to set limits to politics.

It would be unwarranted at this point to imply that the decline in the social authority of science or the erosion of public culture in the liberal-democratic polity is an irreversible trend. It is equally unwarranted to suggest that these developments have already resulted in a radical cultural rejection of the values of science and the Enlightenment. Even as the imperatives of the ethos of science and the canons of scientific rationality are on the defensive, the liberal-democratic society remains under the spell of the Enlightenment and its images of culture and authority. Much of the political appeal of the authority of science persists despite the decline in the willingness to accept science as a model for the perfect fusion of democratic and rational values. The scientist who is no longer accepted on trust as the authentic voice of the public interest is still *qua* scientist regarded as the most informed and, at times, unbiased voice among the partisans. In the liberal-democractic society, governments are judged not only with reference to the adequacy of their "representativeness" but also with regard to their competence and scientific and technical standards for the adequacy of their "performance." The cult of information and facts is still a tenet of the political culture and the commitment to their diffusion persists as a requirement of rational policy-making and the right of the individual citizen to know and to exercise his autonomous judgment.

Like the transformation of religious or national holidays into a collectively coordinated calendar of vacations, the "secularization" of the ethos of science in the civil society involves endless distortions and concessions of norms and values. The resulting attitudes and orientations, although remote from the original cultural code of science,[68] remain significant. Just as the religious values which were incorporated into civic culture survived the decline in the institutional seat of religious authority, so principal elements of the theology of experimental science could survive, in their modified forms, the setbacks which erode the institutional foundations of the temples of empirical reason.

REFERENCES

1. SPRAT, T. (1667) 1958. *History of the Royal Society*. J. Cope and H. W. Jones (Eds.) Washington University Press, St. Louis. Routledge and Kegan Paul Ltd., London.
2. On the role of criticism in science see for instance, K. R. POPPER, *Conjectures and Refutations: The Growth of Scientific Knowledge*. 1968. Harper Torchbooks, New York.
3. MILL, J. S. 1973. On Liberty. *In: The Utilitarians.:* 510. Anchor Books, Garden City, New York.
4. MILL, J. S. Ref. 3: 511.
5. POLANYI, M. The Republic of Science: Its Political and Economic Theory, Autumn, 1962 *Minerva*. **1:** 54.
6. RATNER, J. (Ed). 1939. Science and the Future of Society. *In: Intelligence in the Modern World:* 360. The Modern Library, New York, N.Y.
7. DEWEY J. 1939. *Freedom and Culture:* 148–49. Putnam. New York.
8. DEWEY J. 1963. *Liberalism and Social Action:* 71. Capricorn Books, New York.
9. In the *Journal of Legal and Political Sociology*. 1942. The same article was republished in Merton's *Social Structure and Social Theory* under the title "Science and Democratic Social Structure" and in Merton's *The Sociology of Science* edited by Norman W. Storer as "The Normative Structure of Science." For purposes of citation I have used the last version.
10. Originally published in *Philosophy of Science*. **5** (1938): 321–37. The quotations in the present paper are taken from the version republished in *The Sociology of Science*.
11. MERTON R. K. The Normative Structure of Science: 270.
12. SPRAT T. Ref. 1: 73
13. SPRAT, T. Ref 1: 427.
14. MERTON R. K. Ref. 11: 273.
15. MERTON, R. K. Ref. 11: 274.
16. SPRAT, T. Ref. 1: 85.
17. SPRAT, T. Ref. 1: 75.
18. MERTON, R. K. Ref. 11: 276.
19. MERTON, R. K. Ref. 11: 276.
20. SPRAT, T. Ref. 1: 34.
21. SPRAT, T. Ref. 1: 62.
22. SPRAT, T. Ref. 1: 73.
23. MERTON R. K. Ref. 11: 277–278; and Ref. 10: 264–6.
24. SPRAT, T. Ref. 1: 62.
25. SPRAT, T. Ref. 1: 68.
26. SPRAT, T. Ref. 1: 351.
27. SPRAT, T. Ref. 1: 371.
28. SPRAT, T. Ref. 1: 85.
29. SPRAT, T. Ref. 1: 107.
30. SPRAT, T. Ref. 1: 62.
31. SPRAT, T. Ref. 1: 62.
32. SPRAT, T. Ref. 1: 107.
33. In using the term "cultural code" I am indebted to the extensive works of Shmuel N. Eisenstadt on "Cultural Codes" although in our discussion, the term is not loaded with the same theoretical content given to it in Eisenstadt's work.

34. SPEDDING, J. & R. L. ELLIS, (Eds.) Francis Bacon, De Augmentis, 1875. *Works of Francis Bacon.* Book VI, Ch. II, Vol. IV: 449.

35. JARDINE, L. 1979. *Francis Bacon: Discovery and the Art of Discourse:* 15. Cambridge University Press, Cambridge. Lisa Jardine's very instructive discussion of Bacon has influenced my interpretation of Bacon as did Paolo Rossi's *Francis Bacon.* 1968. S. Rabinovitch (trans.), Routledge & Kegan Paul, London.

36. Robert K. Merton and Paul K. Hatt echoed Bacon's observation concerning the persuasive power of works when they observed that

> "among the greater part of the lay public it would seem that the authority of science rests rather heavily on the technological achievements which it ultimately has made possible. . . These technological accomplishments may also testify to the integrity of the scientist and the power of his research since abstract and difficult theories which cannot be understood or evaluated by the facts are presumably 'proved' in a fashion which can be understood by all through their technological applications. . ."

See their "Election polling Forecasts and Public Images of Social Science," Summer, 1949. *The Public Opinion Quarterly,* 13: 187–188.

37. BACON, F. *The New Organon Book.* I, CXXIV.

38. The English version of Galileo's letter is published in Stillman Drake's *Discoveries and Opinions of Galileo:* 175.

39. MILL, J. S. Ref. 3: 510–511.

40. MILL, J. S. Ref. 3: 510. Charles Webster points out that already in the mid-17th century, "mathematics was thought to render men solitary." See 1975. *The Great Instauration:* 209. Duckworth, London.

41. The mathematical tradition in science and the cultural image of mathematical knowledge as a knowledge based on radical individual autonomy and minimal interdependence and collaboration have apparently had more positive value in non Anglo-American democracies like France. In France, attitudes towards authority appear to involve a peculiar association of freedom with *atomistic individualism.* By contrast to the Anglo-American tendency for *cooperative individualism,* the French conception appears to lead to reliance on public bureaucracy rather than on voluntary communities in constructing and maintaining the public order. Michel Crozier, who examined the bureaucratic phenomenon in modern France, observed that although

> "the privileges and particularisms of the ancient regime have gone. . . the same patterns—individual isolation and lack of constructive co-operative activities on the one side, strata isolation and lack of communication between people of different rank on the other—have persisted."

See his, *The Bureaucratic Phenomenon,* The University of Chicago Press, Chicago, 1964, p. 218. See in this connection also Stanley Hoffmann, "Paradoxes of the French Political Community", and Jesse R. Pitts, "Continuity and Change in Bourgeois France" *In: In Search of France.* Stanley Hoffmann *et al.* (Eds.), Harper Torchbooks, New York, 1963. I would not like as yet to suggest that there is enough evidence to support a strong claim of a definite association between mathematics and French images of knowledge and authority, nor about the differences between the latter and the Anglo-American appropriation of the experimental sciences as a symbol in politics. But even before this line of inquiry is pursued any further, there is, I think, sufficient evidence to suggest the plausibility of these claims.

42. HOBBES, T. 1962. *Leviathan.* M. Oakshott, Ed.: 57. Introduction by Richard S. Peters. Collier Books, New York.

43. HOBBES, T. Ref. 42: 84.

44. MERTON, R. K. Ref. 11: 273–277.

45. SPRAT, T. Ref 1: 62.

46. SPRAT, T. Ref. 1: 426.

47. SPRAT, T. Ref. 1: 73.

48. POLANYI, M. Autumn 1962. "The Republic of Science," *Minerva,* **1:** 54–73; DEWEY, J. 1939. *Freedom and Culture.* Putnam, New York. DEWEY, J. 1939. "Science and the Future of Society" *IN: Intelligence in the Modern World.* J. Ratner Ed.: 343–364. The

Modern Library, New York. Parsons, T. & Platt, G. 1973. *The American University.* Harvard University Press, Cambridge, Mass.

49. SPRAT, T. Ref. 1: 409.

50. TONNIES, F. *Community and Society.* 1963. C. Loomis. Ed. and Trans. Harper Torchbooks, New York.

51. MERTON, R. K. Ref. 10: 258–259.

52. MERTON, R. K. Ref. 10: 265.

53. MERTON, R. K. Ref. 10: 265–266.

54. See, for instance. Leszek Kolakowski, *The Alienation of Reason, A History of Positivist Thought.* 1968. N. Gutterman (trans.), Doubleday, Garden City, New York.

55. FEYERABEND, P. 1970. Against Method. *Minnesota Studies in the Philosophy of Science.*

56. KUHN, T. 1967. *The Structure of Scientific Revolutions.* Phoenix Books, The University of Chicago Press, Chicago.

57. See the apt observations on this point by Stephen Toulmin in: "From Form to Function: Philosophy and History of Science in the 1950s and Now", *Daedalus.* Vol. I, Summer, 1977: 152–161.

58. GREENBERG, D. S. 1967. *The Politics of Pure Science.* New American Library, New York.

59. WATSON, J. D. 1968. *The Double Helix.* Atheneum, New York.

60. SOLOMON, J. J. 1973. *Science and Politics.* The MIT Press, Cambridge, Mass.

61. ROSZAK, T. 1969. *The Making of a Counterculture.* Anchor Books, Garden City, New York.

62. ROSZAK, T. Ref. 61: 323.

63. BLAKE, W. 1976. *Complete Writings.* G. Keynes, Ed.: 617. Oxford University Press, Oxford.

64. See for instance, Alice Kimball Smith, *A Peril and a Hope: The Scientific Movement in America 1945–47.* The University of Chicago Press, Chicago 1965. On the AD-X2 Battery additive controversy, the Test Ben Treaty debate and the controversy over the Salk Vaccine as well as other such controversies, see, "Technical Information for Congress," Report of the Committee on Science and Astranautics, U.S. House of Representatives, 91 Congress, First Session, Library of Congress Serial A, Revised Version, April 15, 1971, U.S. Government Printing Office, Washington, 1971. On the IQ controversy, see Lee J. Cronbach, "Five Decades of Public Controversy Over Mental Testing" and Yaron Ezrahi, "The Jenson Controversy" *In: Controversies and Decisions.* 1976. Charles Frankel, Ed. Russell Sage Foundation, New York.

65. SPRAT, T. Ref. 1:483.

66. See for instance, *Criticism and the Growth of Knowledge.* Imre Lakatos and Alan Musgrave, Eds. Cambridge University Press, Cambridge, 1970. Joseph Ben-David has pointed out other important influences which have contributed to the declining faith in the values of science. Most importantly they include the difficulties which stem from the uses of science to advance military power and the corrosive effects of the growing dependence of science upon government funding. See, *The Scientists' Role in Society,* Prentice-Hall, Inc., Englewood Cliffs, New Jersey, 1971, especially Chapter 9: 169–185.

67. LOCKE, J. 1955. *A Letter Concerning Toleration:* 39. The Library of Liberal Arts, Bobbs-Merrill, Indianapolis.

68. For recent considerations of the strains within and changes of the code of Science, see André Cournand and Harriet Zuckerman 1970. "The code of Science: Analysis and some Reflections on its future," *Studium Generale,* **23:** 141; and André Cournand and Michael Meyer, 1976. "The Scientist's Code," *Minerva,* **14:** 79.

CITATION MEASURES OF THE INFLUENCE OF ROBERT K. MERTON

Eugene Garfield

Institute for Scientific Information
Philadelphia, Pennsylvania 19106

INTRODUCTION

I am deeply honored to contribute to a volume celebrating Robert K. Merton, whose enormous achievements in sociology, the sociology of science, and related areas are so highly and widely esteemed. My personal friendship with Bob Merton extends back about 15 years. On numerous occasions I've acknowledged his moral, intellectual and personal support in developing the *Science Citation Index®* (*SCI®*), and especially, the *Social Sciences Citation Index®* (*SSCI®*), and the *Arts & Humanities Citation Index®* (*A&HCI®*). He serves on the editorial advisory boards of all three of these services. Recently, in one of my essays in *Current Contents®* (*CC®*), I had occasion to reflect upon his rather special place in science.

> I have always had the kind of reaction to much of Merton's writing that I associate with a great novelist, not a great scientist. So much of what he says is so beautifully obvious—so transparently true—that one can't imagine why no one else has bothered to point it out. He is a special kind of scientist: forever reminding us of the forest, while describing it tree by tree.[1]

Over the years, it has been my intuitive feeling that Robert Merton's influence extends well beyond the traditional boundaries of sociology. Further, it has also been my belief that the strength of his influence is derived primarily from his theoretical contributions.

For the purpose of this paper I have decided to test these two subjective notions by conducting a citation analysis of Merton's work. The specific objectives of the study are (1) to define how far beyond the bounds of sociology Merton's influence extends and (2) to determine the extent to which this influence derives from his conceptual, or theoretical, work.

MERTON'S INFLUENCE BEYOND SOCIOLOGY

The data for the study were compiled from the *Science Citation Index* and *Social Sciences Citation Index* for the period 1970–1977. The time period chosen was not entirely arbitrary. At the time we began the study, *SSCI* data was available only for this period. If I began the study a few months later, I could have included data from 1966 because we have now processed social-science material that far back. In any case, we limited our use of *SCI* data, which does go back to 1961, so we would have a common time frame for both data bases.

The *SCI* and *SSCI* were searched to compile a bibliography of all papers published in the natural (*SCI*) and social (*SSCI*) sciences that cited work on which Merton was identified as first author. (A first-author search was considered adequate because Merton is the first or sole author on almost all of his publications. Self citations were excluded from the analysis.) From the bibliography, we compiled the number of citing articles from each subject field or discipline. Thus, the data compiled

61

0028–7113/80/0039–0061 $1.75/2 © 1980 NYAS

TABLE 1

SUMMARY OF DISTRIBUTION OF ARTICLES CITING MERTON FROM 1970–1977

Science Area of Citing Journals	Number of Articles Citing Merton	Percent of Total
Natural Sciences	203	8
Sociology	925	36
Social Sciences (Excluding Sociology)	1413	56
Total	2541	100.0

reflects not the number of individual citations to Merton, but the number of articles citing Merton by authors other than himself. Fields were defined by the subject-classification assigned to journals in *SCI*[2] or *SSCI*.[3] For this study the "discipline" of the citing article was considered the same as that for the journal in which it was published.

TABLE 1 summarizes the basic findings of the survey of articles citing Merton in the years 1970–1977. The total number of 2541 citing articles, in itself, is an indication of Merton's impact. Their distribution over the three general categories of the natural sciences, sociology, and the social sciences excluding sociology are equally revealing. The two categories outside of sociology, the natural sciences and the social sciences other than sociology, account for 1616 citing articles—64% of the total.

The figures in the table acquire even stronger significance when compared with the average annual citation frequencies of other authors cited in the natural and social sciences. In the natural sciences, the average annual figure over the eight years studied was 7.05;[4] in the social sciences, it was 3.48.[5] If one simply multiplied these figures by 8 to cover the number of years studied, you would obtain a summed average of 56.4 in the natural sciences and 27.84 in the social sciences. These figures imply, of course, that the average author is cited every year, which certainly is not the case. In that sense, they are a very conservative benchmark for measuring Merton's relative impact.

Nevertheless, even when the 925 citations from sociology are excluded, the remaining 1413 social science citations (TABLE 1) are over 50 times greater than average and over 80 times greater if sociology is included. And even in the natural sciences, the 203 citations of his work is almost four times the average of 56.4 for a natural scientist.

ANALYSIS BY SOCIAL SCIENCE DISCIPLINE

TABLE 2 shows in more detail the pattern of Merton's influence in the social sciences. In that table, the citations he received between 1970–1977 are distributed by the disciplines of the citing journals.

Predictably, the largest category, with 40% of the total, is Sociology, which includes the sub-specialty of demography. (Demography articles account for less than 5% of the category). The other categories exhibit a relatively even distribution of citations, ranging from a high of 8% for Education and Political Science to a low of 1% for Theology.

The Miscellaneous Journals category (7%) is a catchall for journals covering fields too small to be included as separate entries. This category includes communications, social research, linguistics, urban studies, ethnic studies, and industrial relations. The broad diversity of this category is a further indication of the wide range of Merton's impact.

Next in importance are three categories, which each account for a 5% share of the citations: Business and Economics, History and History of Science, and Psychology. The Business and Economics category comprises business administration, economics, and economic history journals. Merton is cited here primarily for his work on bureaucracy and personality structure, along with his work on technological emergence as it relates to the industrial revolution.

Merton's Ph.D. dissertation on "Science, Technology, and Society in 17th Century England" is especially interesting. It has been a strongly influential work in the study of the history of science. Of course, Merton's prolific work in the sociology of science has also had impact on historians of science.

Merton's impact on psychology derives largely from his interest in the effects of certain social structures (e.g., bureaucracies) on individual psychosocial tendencies.

Next in rank, the Interdisciplinary Social Science journals (4%) consist of those publications not readily classifiable as belonging to a single traditional social science discipline. Included are *Policy Science, Development and Change, Journal of Development Studies,* and *General Systems.* This group of citations attests further to the multidisciplinary impact of Merton's work.

The showing of the remaining categories, each of which accounts for less than 5% of the citations in the social sciences, is more a reflection of small numbers than of little impact. Certain fields like anthropology produce less literature than sociology or psychology. In relation to the size of the anthropology literature, 39% is not inconsequential. In fact, the 78 articles from anthropology suggest that Merton's work on functional analysis has had considerable impact.

In other cases, such as social psychology, the small numbers are due to the

TABLE 2

DISTRIBUTION, BY DISCIPLINE, OF SOCIAL-SCIENCE ARTICLES CITING
MERTON FROM 1970–1977

Discipline of Citing Journals	Number of Articles Citing Merton	Percent of Total
Anthropology	78	3
Business and Economics	127	5
Education	177	8
History and History of Science	124	5
Interdisciplinary Social-Science Journals	100	4
Law	55	2
Philosophy and Philosophy of Science	45	2
Political Science	190	8
Psychology	127	5
Sociology and Demography	925	40
Social Issues	69	3
Social Psychology	44	2
Social Work	65	3
Theology	26	1
Miscellaneous Journals	167	7
Multidisciplinary Journals	19	1
Total	2338	99

classification system we adopted. The citing articles were categorized by the specialty of the publishing journal. There are relatively few journals dedicated to social psychology. Much of the literature of this field is published in sociology and psychology journals.

In the cases of philosophy, the philosophy of science, and theology, the small numbers can be attributed to the ambiguous position of these disciplines within the social sciences. Philosophy, in all its variations, has come to be more of a discipline of the humanities than of the social sciences. The coverage of *SSCI* reflects this perspective; its coverage of 48 philosophy journals amounts to only 65% of the coverage provided by the *Arts & Humanities Citation Index (A&HCI)*.[6] The case is even more clear-cut in theology, where the *SSCI* covers 8 journals compared to the 57 covered by *A&HCI*. (*A&HCI* was not used in the study because it only covered two years at the time.)

Some of the statistics in these categories, however, are surprising even within the framework of the small numbers involved. Considering that Merton's work on deviant behavior, particularly his contribution to anomie theory, is widely taught and used in social work, one would have expected the literature of that field to cite him more heavily. The fact that it doesn't might reflect another of his concepts: obliteration by incorporation into the common knowledge.[7]

Conversely, the number of citing articles from law seems extraordinarily high for a sociologist, as does the number from theology. Since it is primarily his early work that is relevant to these fields, these findings suggest not only deep impact but also unusual longevity.

The multidisciplinary-journal category consists of only those few journals that cover all the sciences, social and natural alike, such as *Science* and *Nature*. Since they are not primarily social science journals, we included only those citing articles that were clearly on social science subjects. Articles from multidisciplinary journals categorized as natural science are covered in the next section.

ANALYSIS BY NATURAL SCIENCES DISCIPLINE

Turning now to the findings in the natural sciences, we see from TABLE 3 that Merton's impact certainly ranges well beyond the social sciences. (It should be noted that the term, natural sciences, is being used to denote all disciplines beyond the social sciences.)

As TABLE 3 shows, the largest number of articles citing Merton occurs in the field of medicine. Many of these articles concern his work in *The Student-Physician*[8] and, to a somewhat lesser extent, his *Social Theory and Social Structure*.[9] In general, the medical-journal articles citing Merton discuss aspects of social medicine or the education of doctors and other medical personnel.

Psychiatry, with 31% of the citations, ranks as the second largest category. Psychiatry, of course, has a strong social-science orientation. But since it is also a medical specialty we treated it as part of the natural sciences. In contrast to the medical category, *Social Theory and Social Structure* is cited much more frequently by psychiatrists than *Student-Physician*.

Information science (19%) is the third largest category. Merton's relevant contribution to this field comes from his examination of the growth and structure of science, and from his work in information exchange among scientists.

The relatively small number of citing articles in the physical and biological sciences is, of course, quite understandable, since these areas are the most remote from Merton's own field. Some of the articles in biology and the physical sciences concern

TABLE 3

DISTRIBUTION, BY DISCIPLINE, OF NATURAL-SCIENCE ARTICLES CITING
MERTON FROM 1970–1977

Discipline of Citing Journals	Number of Articles Citing Merton	Percent of Total
Biology and Biochemistry	10	5
Information Science	38	19
Medicine	75	37
Multidisciplinary Journals	6	3
Physical Science	12	6
Psychiatry	62	30
Total	203	100

the design of courses in their respective fields for use in liberal arts curricula. Other articles discuss historical analyses of major cognitive events, theories and methodologies within the fields of the authors. Merton is cited principally for his work on the social context of such developments.

As noted previously, Merton's citation record is almost 400% greater than that of the average cited author in the natural sciences. This prompted me to extend the analysis back to 1961 to see whether the pattern holds. TABLE 4 presents the results of that more extensive analysis. Adding nine years to the study almost doubled the number of articles citing his work. But the distribution of citing articles over the natural science disciplines remains essentially the same. The largest difference in citation levels is a 6% increase in the information science category for the most recent years. This reflects the increasing contact between information scientists and people working in the sociology of science.

MERTON'S CONTRIBUTION: CONCEPTUAL OR EMPIRICAL?

With the breadth of Merton's influence beyond sociology clearly confirmed, the focus of the study shifted to several questions about the nature of his influence. Have scientists outside of sociology used Merton's work more for his empirical observations

TABLE 4

DISTRIBUTION, BY DISCIPLINE, OF NATURAL-SCIENCE ARTICLES CITING
MERTON FROM 1961–1977

Discipline of Citing Journals	Number of Articles Citing Merton	Percent of Total
Biology and Biochemistry	15	4
Information Science	49	13
Medicine	147	40
Multidisciplinary Journals	10	3
Physical Science	21	6
Psychiatry	123	34
Total	365	100

(findings) or for his concepts? What cognitive areas of his work are cited in the nonsocial sciences, and do these differ from the areas cited in the social sciences? Are the concepts for which Merton is cited distinctively Mertonian, or do they have a more general origin?

To explore these questions, a content analysis was done on 35 social science and 49 natural science articles published between 1961–1977 to determine the nature of Merton's work cited by authors in disciplines other than sociology. More natural-science articles were examined because they are so far beyond the usual sphere of influence of a sociologist.

The sample of journals was dictated by journal availability in local libraries rather than by statistical standards. Nevertheless, the articles were randomly selected within that constraint. It must be mentioned, however, that the content analysis employed here is far from the rigorous and systematic content analysis used by Cole[10] in his study of Merton's citations within sociology. Rather, the content analysis that follows should be viewed as illustrative.

TABLE 5 looks at whether Merton was cited more frequently for concepts or for findings. A "Secondary Citations" category was included here for cases in which an author cited a Merton article for something that Merton, himself, cites or quotes (e.g. W. I. Thomas or St. Augustine). Also included in this category are articles that cited Merton only in the bibliography, not in the text.

It is worth repeating that none of the articles analyzed for content are from sociology journals as such. The Social Sciences category in this table (and in those that appear later) refers to social sciences other than sociology.

From the date in TABLE 5 we see that 63% of the inspected citations in the natural sciences are for conceptual contributions while in the social sciences it is 83%. In both the natural and social sciences, then, at least two-thirds of the citations are for concepts rather than findings. This confirms the impression that Merton's major contribution has been that of a theorist.

CONCEPTS OF INFLUENCE

It is interesting to examine the specific concepts for which Merton has been cited in the natural and social sciences. Such an examination shows that his breadth of influence stems from an equally broad range of conceptual contributions. There were 31 conceptual citations in the natural science articles and 29 in the social science articles. These citations were to 26 different concepts, 10 of which were common to

TABLE 5

DISTRIBUTION OF MERTON CITATIONS BY NATURE OF MATERIAL CITED

Science Area of Citing Journals	Concepts		Findings		Secondary Citations		Total	
	Number*	%†	Number	%†	Number	%†	Number	%†
Natural Sciences	31	63	12	24	6	12	49	99
Social Sciences (Excluding Sociology)	29	83	5	14	3	1	35	100
All Science (Excluding Sociology)	60	71	17	20	7	8	84	99

*Number of Citing Articles.
†Percent of Total.

TABLE 6

DISTRIBUTION OF CONCEPTUAL CITATIONS TO MERTON BY
COGNITIVE AREA OF CITED CONCEPT

Cognitive Area of Cited Concept	Science Area of Citing Journals			
	Natural Sciences		Social Sciences	
	Number*	%†	Number*	%†
Functional Analysis	9	29	12	42
Deviance	7	23	9	31
Sociology of Science	8	26	3	10
Professional Socialization	5	16	0	0
Other	2	6	5	17
Total	31	100	29	100

*Number of Citing Articles.
†Percent of Total.

both groups of citing articles. The natural-science articles cited an additional 10 concepts while the social-science articles cited another six.

Within this very wide conceptual range, three specific concepts tend to dominate. In the social sciences, 43% of the conceptual citations were to Merton's work on manifest and latent functions (20%), social structure and anomie (13%), and reference group theory (10%). Two of those concepts also account for 22% of the citations from the natural sciences: social structure and anomie (14%) and manifest and latent functions (8%).

Since many of the individual concepts for which Merton is cited fall into a few broad cognitive areas, it is possible to collapse the data to obtain a clearer view of the sources of his conceptual influence. TABLE 6 presents such a view. The main cognitive areas of Merton's conceptual work are listed along with the number of social- and natural-science papers that cited them. Concepts were assigned to cognitive areas as follows:

Functional Analysis: functional analysis in general; manifest and latent functions; role sets; general concept of anomie.

Deviance: social structure and anomie; conformity in group size; alienation; bureaucracy.

Sociology of Science: priority disputes; Matthew effect; information structure of science; multiple discoveries; norms of science.

Professional Socialization: anticipatory socialization; professional socialization; norms of medical culture; informal medical curricula.

Other: concepts which could not be fitted readily into the other categories, such as sex typing, self-fulfilling prophecy, discussions of Mannheim, middle-range theory, analysis of political machines.

TABLE 6 shows approximately equal levels of influence for the concepts of functional analysis (29%), sociology of science (26%), and deviance (23%) in the natural sciences. Professional socialization concepts, the target of 16% of the citing natural science papers, are not far behind.

The pattern of conceptual influences in the social sciences is considerably different. The concepts of functional analysis and deviance are clearly more dominant in the social sciences than in the natural sciences. The 31% figure for the concepts of deviance indicate that Merton's work in this area remains important despite Cole's observation of a shift in emphasis in this literature to symbolic interaction.[10]

Conversely, the concepts in the sociology of science and professional socialization are considerably less important in the social sciences than they are in the natural sciences. Authors in the natural sciences tend to cite Merton's work in the sociology of science in connection with studies of historical developments in the field; he usually is cited to relate their analyses to more general principles of scientific development. The citation potential for these concepts in the social sciences is limited to the relatively few people working in the sociology of science specialty. The relatively low frequency with which they are cited reflect the relative position of that specialty within the social sciences.

None of the social-science articles examined cited Merton's concepts on professional socialization. This is in sharp contrast to the relatively strong influence of these concepts in the natural sciences. One explanation is that Merton's work in this area is exerting more influence at the practical level of curriculum development than at the theoretical level of sociological research. And, in fact, the natural-science papers that cite this particular set of concepts do so in discussions of discipline-specific educational curricula. Another, more likely, explanation is that the initial impact of these concepts upon the social sciences was felt in the field of sociology in the 1950s, when they first appeared. If so, this part of the study, which was designed to look at the nature of Merton's influence beyond sociology, and went back only through the 1960s, would show no signs of the impact.

Another point of impact that doesn't show up is Merton's concept of the self-fulfilling prophecy. One of the concepts in the *Other* category, the infrequency with which it is cited belies its widespread use. As so many concepts whose influence is particularly pervasive, Merton's self-fulfilling prophecy seems to have suffered obliteration by being incorporated into the general body of knowledge that is considered common property.

ROOTS OF CONCEPTS

The other aspect of Merton's influence that was examined has to do with the roots of his cited concepts. Are the concepts for which he is generally cited ones that are original to him or are they ideas of more general origin?

Since there is a certain element of interpretive subjectivity in this type of analysis, some examples are needed to demonstrate the types of judgments made. Typical of citations to distinctively Mertonian concepts are:

Merton's (1949) seminal essay on manifest and latent functions made it clear that intended and known functions (goals) frequently carry different consequences for a social system than unintended and unknown functions.[11]

This kind of consideration can be used to extend Merton's classification of type of adaption to society.[12]

. . . self-fulfilling or self-frustrating prophecies. Robert K. Merton has analyzed these predictions from a sociological perspective . . .[13]

Robert Merton . . . sees the priority disputes as "signposts announcing the violation of the social norms" of the scientific establishment.[14]

Merton offers the theory that anomie occurs as a result of discrepancy between culturally shared goals and the means for achieving them.[15]

. . . we are concerned with conflicting objectives which occur as "unanticipated consequencies". . .[16]

Typical of citations judged to be more general concepts discussed by Merton are:

The concept of anomie or "normlessness," for example, has had a profound effect on modern sociology and psychology.[17]

The high delinquency rates often found in lower-class city slums and among minority groups (Sutherland & Cressey, 1966) have been used as the basis for theories attributing delinquency to lower-class mores, social disorganization, and culture conflict (cf. Cloward & Ohlin, 1960; Cohen, 1955; Glaser, 1965; Merton, 1957; Miller, 1958; Sellin, 1938).[18]

Gurr has proposed that social discontent is most often produced by decremental economies. Perhaps the decremental curve is most effective because people tend not to compare themselves to dissimilar others (Berkowitz, 1972; Festinger, 1954; Merton & Kitt, 1950).[19]

Temperamental differences contribute their additions to these barriers. Obviously, we do not deliberately choose the differences in our temperaments. Some of these temperamental differences have been pointed out by Roe, Ginzburg, Ginzburg, Axelrad and Herma, Eiduson and Merton, to mention a few.[20]

... a state of anomie. This concept has been defined as a "condition of relative normlessness in a society or group" ... (Merton, 1957).[21]

... the social nature of chemical discovery, stems from the work of such sociologists of science as Merton and Storer.[22]

TABLE 7

ROOTS OF CITED MERTON CONCEPTS

Concept Origin	Science Area of Citing Journals			
	Natural Sciences		Social Sciences	
	Number*	%†	Number*	%†
Merton	20	65	22	76
General	11	35	7	24
Total	31	100	29	100

*Number of Citing Articles.
†Percent of Total.

These examples show that the analysis was discriminating enough to distinguish between a citation to a general Merton discussion of anomie, a concept that did not originate with him, and a citation to a Merton discussion of anomie that attributes the effect to a disparity between goals and the means of achieving them, which is a unique Merton extension of the concept.

The results of the analysis are shown in TABLE 7. Though the percentage of Merton concepts cited by natural-science articles is 11% lower than the percentage cited by social-science articles, Merton's own concepts accounted for substantially more than 50% of the citations in both classes of articles. These data make it clear that Merton's influence is primarily the result of work unique to him, even in the natural sciences, which are the most remote from his own discipline.

CITED WORKS

To round out the study, we decided to complement the analysis of the fields citing Merton with an analysis of the work they have been citing. TABLE 8 identifies the Merton publications that were cited in *SSCI* five or more times during the period

TABLE 8

MERTON PUBLICATIONS CITED IN SSCI®FIVE OR MORE TIMES FROM 1969–1977

Publication	Citations
Merton, R. K. *Social Theory and Social Structure.* (New York: The Free Press, 1968). 702 pp.	1418
—, Reader, G., Kendall, P. L. (eds.). *The Student-Physician: Introductory Studies in the Sociology of Medical Education.* (Cambridge: Harvard University Press, 1957). 360 pp.	93
—, Nisbet, R. A. (eds.) *Contemporary Social Problems.* (New York: Harcourt Brace Jovanovich, 1961). 754 pp.	79
—. "Social Structure and Anomie." *Amer. Sociological Review* 3:672, 1938.	60
—, Broom, L., Cottrell, L. S. Jr. (eds.) *Sociology Today: Problems and Prospects.* (New York: Basic Books, 1959). 623 pp.	57
—. *The Sociology of Science: Theoretical and Empirical Investigations.* (N. Storer, ed.) (Chicago: University of Chicago Press, 1973). 605 pp.	57
—. "Priorities in Scientific Discovery." *Amer. Sociological Review* 22:635, 1957.	51
–. "The Matthew Effect in Science." *Science* 159:56, 1968.	49
—. Science, Technology and Society in Seventeenth Century England. In (G. Sarton, ed.) *OSIRIS: Studies on the History of Learning and Culture.* (Belgium: The St. Catherine Press, 1938). p. 362–632.	38
—. "Bureaucratic Structure and Personality." *Social Forces* 18:560, 1939.	35
—, Gray, A., Hockey, B., Selvin, H. (eds.) *Reader in Bureaucracy.* (New York: The Free Press, 1952). 464 pp.	31
—. "The Role-Set: Problems in Sociological Theory." *Brit. J. Soc.* 8:106, 1957.	31
—. "The Self-fulfilling Prophecy." *Antioch. Rev.* 8:193, 1948.	29
—. *On Theoretical Sociology: Five Essays, Old and New.* (New York: The Free Press, 1967). 180 pp.	28
–, Lazarsfeld, P. F. (eds.) *Continuities in Social Research: Studies in the Scope and Method of "The American Soldier."* (New York: The Free Press, 1950). 255 pp.	26
—. "The Unanticipated Consequences of Purposive Social Action." *Amer. Sociological Review* 1:894, 1936.	23
—. "Singletons and Multiples in Scientific Discovery: A Chapter in the Sociology of Science." *P. Am. Philos. Soc.* 105:470, 1961.	21
—. "Behavior Patterns of Scientists." *Amer. Sci.* 57:1, 1969.	18
—. "Insiders and Outsiders: A Chapter in the Sociology of Knowledge." *Amer. Journal of Sociology* 78:9, 1972.	16
—, Barber, E. "Sociological Ambivalence." In: (Edward Tiryakian, ed.) *Sociological Theory, Values and Sociocultural Change.* (New York: The Free Press, 1963). p. 91–120.	16
—. "Intermarriage and Social Structure." *Psychiatry* 4:361, 1941.	15
—. "Anomie, Anomia and Social Interaction: Contexts of Deviant Behavior. "In: (Marshall Clinard, ed.) *Anomie and Deviant Behavior.* (New York: The Free Press, 1964) p. 213–42.	13
—, Fiske, M., Kendall, P. *The Focused Interview.* (New York: The Free Press, 1956). 186 pp.	12
—. "Resistance to the Systematic Study of Multiple Discoveries in Science." *Arch. Eur. Sociol.* 4:237, 1963.	12
—. "The Role of Applied Science in the Formulation of Policy." *Philos. Sci.* 16:161, 1949.	11
—. "Social Conformity, Deviation, and Opportunity-Structures." *Amer. Sociological Review* 24:177, 1959.	10
—. *Mass Persuasion.* (New York: Harper & Brothers, 1946). 210 pp.	8
—. "Patterns of Influence: A Study of Interpersonal Influence and Communications Behavior in a Local Community." In: (Paul Lazarsfeld &	

TABLE 8 (*Continued*)

Publication	Citations
Frank Stanton, eds.) *Communications in Research.* (New York: Harper & Brothers, 1948–49). p. 180–219.	7
_. "Civilization and Culture." *Sociology and Social Research* 21:103, 1936.	5
_. "Discrimination and the American Creed." In: (R. M. McIver, ed.) *Discrimination and National Welfare.* (New York: Harper & Brothers, 1948). p. 99–126.	5
–. "Social Psychology of Housing." In: (W. Dennis, ed.) *Current Trends in Social Psychology.* (Pittsburgh: University of Pittsburgh Press, 1948). p. 163–217	5

1969–1977, and shows how the citations were distributed. The year 1969 was included in this analysis, because the data was taken from an existing compilation of highly-cited articles in *SSCI* for the 1969–1977 time period. TABLE 9 identifies all Merton publications that were cited in *SCI* during the same time period, and also shows how the citations were distributed. The citation frequencies shown in the two tables are not mutually exclusive, because we made no attempt to correct for the degree of overlap in the coverage patterns of *SCI* and *SSCI*. The two tables, therefore are not directly comparable in any way.

Nevertheless, the tables do clearly show that the books Merton wrote and edited have been, by far, the major source of his influence. They accounted for 81% of his *SSCI* citations and 76% of those from *SCI*. *Social Theory and Social Structure,* which probably contains the broadest mix of his observations and ideas, is, appropriately, the most heavily cited of his writings. It accounts for 62% of his *SSCI* citations and 57% of those from *SCI*.

CONCLUSION

Not surprisingly, then, the data emphatically confirm my intuitive judgment that the influence of Robert K. Merton ranges far beyond his home discipline of sociology. Not only is he highly cited throughout the social sciences, but the pattern of citations to his work reveal he has had a considerable impact in the natural sciences as well. The data also show that the main strength of his influence rests on his prolific production of unique sociological concepts that are widely accepted and used. However, his own research "findings" have also been widely used by others in a variety of contexts.

Besides confirming these intuitive judgments, the study results also suggest directions for additional research that would provide a more definitive picture of Merton's influence. An underlying assumption of this study and all others that make comparisons based on citation statistics is that all citations are equal. There are at least two reasons for considering that assumption to be an oversimplification.

The first is that the probability of being cited in a given field is affected by the size of the literature and the average number of references per article in the field. Since these factors vary from field to field, so does citation potential. To develop a more accurate view of Merton's influence, the frequency with which he is cited in each field should be weighted by the citation potential of the field.

The other way in which straight citation counts oversimplify reality is by failing to say anything about why authors cite someone. We scratched the surface of this

TABLE 9

MERTON PUBLICATIONS CITED IN SCI® FROM 1969–1977

Publication	Citations
Merton, R. K. *Social Theory and Social Structure.* (New York: The Free Press, 1968). 702 pp.	382
—, Reader, G., Kendall, P. L. (eds.). *The Student-Physician: Introductory Studies in the Sociology of Medical Education.* (Cambridge: Harvard University Press, 1957). 360 pp.	57
—. "The Matthew Effect in Science." *Science* 159:56, 1968.	45
—. *The Sociology of Science: Theoretical and Empirical Investigations.* (N. Storer, ed.) (Chicago: University of Chicago Press, 1973). 605 pp.	36
—. "Priorities in Scientific Discovery." *Amer. Sociological Review* 22:635, 1957.	23
—. "Behavior Patterns of Scientists." *Amer. Sci.* 57:1, 1969.	16
—. "Science, Technology and Society in Seventeenth Century England." In (G. Sarton, ed.) *OSIRIS: Studies on the History of Learning and Culture.* (Belgium: The St. Catherine Press, 1938). p. 362–632.	15
—. "The Self-fulfilling Prophecy." *Antioch. Rev.* 8:193, 1948.	14
—. "Singletons and Multiples in Scientific Discovery: A Chapter in the Sociology of Science." *P. Am. Philos. Soc.* 105:470, 1961.	13
—, Lazarsfeld, P. F. (eds.) *Continuities in Social Research: Studies in the Scope and Method of "The American Soldier."* (New York: The Free Press, 1950).	11
—, Broom, L., Cottrell, L. S. Jr. (eds.) *Sociology Today: Problems and Prospects* (New York: Basic Books, 1959). 623 pp.	11
—, Nisbet, R. A. (eds.) *Contemporary Social Problems.* (New York: Harcourt Brace Jovanovich, 1961). 754 pp.	10
—. "Anomie, Anomia and Social Interaction: Contexts of Deviant Behavior." In: (Marshall Clinard, ed.) *Anomie and Deviant Behavior.* (New York: The Free Press, 1964) p. 213–42.	8
—. "Patterns of Influence: A Study of Interpersonal Influence and Communications Behavior in a Local Community." In: (Paul Lazarsfeld & Frank Stanton, eds.) *Communications in Research* (New York: Harper & Brothers, 1948–49). p. 180–219.	8
—. "Social Structure and Anomie." *Amer. Sociological Review* 3:672, 1938.	8
—, Fiske, M., Kendall, P. *The Focused Interview.* (New York: The Free Press, 1956). 186 pp.	8
—. "The Role-Set: Problems in Sociological Theory." *Brit. J. Soc.* 8:106, 1957.	8
—. *On the Shoulders of Giants: A Shandean Postscript.* (New York: The Free Press, 1965). 289 pp.	7
—, Gray, A., Hockey, B., Selvin, H. (eds.) *Reader in Bureaucracy.* (New York: The Free Press, 1952). 464 pp.	7
—. "Resistance to the Systematic Study of Multiple Discoveries in Science." *Arch. Eur. Sociol.* 4:237, 1963.	7
—, Barber, E. "Sociological Ambivalence." In: (Edward Tiryakian, ed.) *Sociological Theory, Values and Sociocultural Change.* (New York: The Free Press, 1963). p. 91–120.	7
—. "Social Psychology of Housing." In: (W. Dennis, ed.) *Current Trends in Social Psychology.* (Pittsburgh: University of Pittsburgh Press, 1948). p. 163–217	6
—. "The Unanticipated Consequences of Purposive Social Action." *Amer. Sociological Review* 1:894, 1936.	6
—. "The Functions of the Professional Association." *Am. J. Nurs.* 58:50, 1958.	5
—. "Intermarriage and Social Structure." *Psychiatry* 4:361, 1941.	4
—. *Mass Persuasion.* (New York: Harper & Brothers, 1946). 210 pp.	4
—. *On Theoretical Sociology: Five Essays, Old and New.* (New York: The Free Press, 1967). 180 pp.	4

question by analyzing the contents of a sample of articles to determine whether Merton was being cited for his conceptual work or for his empirical findings. S. Cole dug considerably deeper, though across a narrower front, in his analysis of the deviance literature.[10] J. Cole and H. Zuckerman conducted a similar study of the sociology of science literature that added substantially to this line of research.[23] Both of those studies involved an in-depth content analysis of a sample of papers that cited Merton to find out how the citing authors used his work. A similar type of analysis of citations to Merton from outside the field of sociology would elaborate in a useful way on our findings about the nature of his interdisciplinary contributions.

In their work on Merton's influence in the sociology-of-science, Cole and Zuckerman[23] point to another research gap that is particularly pertinent to the subject of measuring the influence of Merton or any other scientist through citation analysis. They wrote: "In the absence of statistical norms on the relative frequency of different kinds of citations in the sociological literature, it is not possible to interpret the distribution observed here." They were commenting on statistics on different kinds of citations, but the observation is equally as relevant to simple citation counts that do not distinguish between the different motives authors may have for citing a given work. Though general averages of citation frequency can be used to provide some rough benchmarks that are useful in interpreting citation counts of individuals, there is an obvious need for more specific benchmarks.

It would be useful to know, for example, how frequently typical authors are cited in each of the scientific disciplines, how frequently influential authors are cited in each of the disciplines, and the average frequency with which influentials are cited from outside their fields. Anyone who provides valid answers to these types of questions will be making a significant contribution to the study of science.

The expert sociologist may well ask how an information scientist can have the hubris to discuss his intuitive judgements about Merton's influence or impact. I cannot even claim to have read all of Bob Merton's prolific work. And certainly I have only read a miniscule portion of the literature which has cited him. But for more than 15 years I have received a weekly computer report which has described for me, by title, author, and journal, every new article added to the ISI® data base that has cited Merton's work and the work of many of his students. As a consequence of examining more than 800 of these ASCA® reports,[24] I feel justified in claiming a certain intuitive expertise about Merton. There is no doubt in my mind, however, that it requires the judgement of a practicing sociologist who is steeped in the literature of sociology to express an opinion on that part of Merton's influence which cannot be measured quantitatively as we have done here. It is such cognitive familiarity with the literature that enables one's peers to estimate the worth of contributions whose sources have long since been obliterated.

References

1. GARFIELD, E. 1977. Robert K. Merton: Among the Giants. *Current Contents,* **28:** 5–7.
2. Institute for Scientific Information. *Science Citation Index, 1977 Annual. Guide and Lists of Source Publications:* 148. Philadelphia. ISI, 1978.
3. —*Social Sciences Citation Index, 1977 Annual. Guide and Journal Lists.* p. 108. Philadelphia. ISI, 1978.
4. —Science Citation Index 1961–1977 Comparative Statistical Summary. *SCI 1977 Annual. Guide and Lists of Source Publications:* 22. Philadelphia, ISI, 1978.
5. —Social Sciences Citation Index 1970–1977 Comparative Statistical Summary. *SSCI 1977 Annual. Guide and Journal Lists:* 20. Philadelphia: ISI, 1978.

6. GARFIELD, E. August 8, 1977. Will ISI's Arts & Humanities Citation Index revolutionize scholarship? *Current Contents* **32:** 5–9.
7. MERTON, R. K. 1968. *Social Theory and Social Structure:* 27–29, 35–38. The Free Press, New York, N.Y.
8. —*The Student-Physician: Introductory Studies in the Sociology of Medical Education.* 1957. R. K. Merton, G. Reader & P. L. Kendall, Eds.: 360. Harvard University Press, Cambridge.
9. MERTON, R. K. 1968. *Social Theory and Social Structure.* 3rd ed. The Free Press, New York, N.Y.
10. COLE, S. 1975. The Growth of Scientific Knowledge: Theories of Deviance as a Case Study. *In: The Idea of Social Structure.* L. A. Coser, Ed. 175–220. Harcourt, Brace and Jovanovich, New York, N.Y.
11. GILLESPIE, D. F., D. S. MILETI, R. E. COTZ & R. W. PERRY. 1976. Historical and Paradigmatic Differences in the Use of the Goal Concept. *Int. Rev. Hist. Polit. Sci.* **13:** 4.
12. GOLDBERG, C. 1973. Some Effects of Fear of Failure in the Academic Setting. *J. Psychol.* **84:** 331.
13. SCHULMAN, P. K. 1976. The Reflexive Organization: On Decisions, Boundaries and the Policy Process. *J. Polit.* **38:** 1022.
14. DUCKWORTH, D. 1976. Who Discovered Bacteriophage? *Bacteriol. Rev.* **40:** 793.
15. BRENNER, M. H. 1975. Trends in Alcohol Consumption and Associated Illnesses. *Am. J. of Publ. Heal.* **65:** 1289.
16. STEPHENSON, R. W. & B. S. GANTZ. 1965. Conflicting Objectives in a Research and Development Organization. *IEEE Transact. Engin. Manag.* **12:** 125.
17. STANNARD, D. E. 1973. Death and Dying in Puritan New England. *Am. Hist. Rev.* **78:** 1328.
18. MEGARGEE, E. I., R. V. LEVINE & G. V. C. PARKER. 1971. Relationship of Familial and Social Factors to Socialization in Middle-class College Students. *J. Abnor. Psychol.* **77:** 76.
19. ROSS, M. & M. J. MCMILLEN. 1973. External Referents and Past Outcomes as Determinants of Social Discontent. *J. Experiment. Soc. Psychol.* **9:** 447.
20. KUBIE, L. S. 1970. Problems of Multidisciplinary Conferences, Research Teams, and Journals. *Perspect. Biol. Med.* **13:** 412.
21. PHOON, W. O., S. R. QUAH, C. Y. TYE & H. K. LEONG. 1976. A Preliminary Study of the Health of a Population Staying in Apartments of Varying Sizes in Singapore. *Annal. Trop. Med. Parasitol.* **70:** 243.
22. FENSHAM, P. 1976. Social Content in Chemistry Courses. *Chem. Brit.* **12:** 148.
23. COLE, J. R. & H. ZUCKERMAN. 1975. The Emergence of a Scientific Specialty: The Self-Exemplifying Case of the Sociology of Science. *In: The Idea of Social Structure.* L. A. Coser, Ed.: 139–174.
24. GARFIELD, E. & I. H. SHER. 1967. *ASCA (Automatic Subject Citation Alert),* a New Personalized Current Awareness Service for Scientists. *Am. Behavior. Scient.* **10:** 29–32.

THE ROLE OF PSYCHOLOGICAL EXPLANATIONS OF THE REJECTION OR ACCEPTANCE OF SCIENTIFIC THEORIES

Adolf Grünbaum

University of Pittsburgh
Department of Philosophy
Pittsburgh, Pennsylvania 15260

In a public lecture at my University, the philosopher Michael Scriven challenged the credentials of psychoanalytic treatment. Immediately afterward, a senior psychoanalyst in the audience turned toward me to inquire whether Scriven's father or brother was an analyst. Evidently, the interlocutor deemed it unnecessary to come to grips with the lecturer's *arguments* for doubting the capability of available clinical evidence to sustain the claims of efficacy that had been made for Freudian psychotherapy.

Another colleague, concerned with the light that psychoanalytic principles might throw on some of the humanistic disciplines, concluded that the purported insights afforded by these principles are largely all-too-facile pseudo-explanations. Professional psychoanalysts present at university lectures in which he expounded this scepticism across the country usually responded rather patronizingly as follows: They offered diagnoses of the neurosis that had allegedly impelled the sceptical colleague to reject psychoanalytic theory after he presumably experienced ego-threat from it. Incidentally, the analysts in question repeatedly offered these dismissive psychological explanations with great confidence, undaunted by the fact that a Freudian diagnosis avowedly requires a considerable number of analytic sessions. Perhaps it is therefore not surprising that no two analysts offered the *same* diagnosis as to the sceptic's presumed neurotic affliction.

Far from being atypical, such psychologistic responses to criticism are, alas, rather representative, as illustrated by a comment on my forthcoming book *Is Psychoanalysis a Pseudo-Science?*, given to me by a practicing psychoanalyst. He pointed out that typical analyst readers will be looking not so much at my reasoning as at those of my psychological motivations purportedly discernible from my citation of derogatory assessments of analysis by others.

Indeed, I encountered the same dismissive psychologism as my sceptical colleague above when I recently examined the credentials of Freudian theory from a philosophy of science perspective in a lecture on a campus in Arizona. And the substantive issues posed in that Arizona encounter will now serve as my point of departure for dealing with the central question of this paper, which is the following: Just what ought to be the role, if any, of giving *psychological* explanations of either the rejection or the acceptance of supposedly scientific theories of man or nature? And similarly for the rejection or acceptance of philosophical beliefs or religious doctrines.

The lecture I gave in Arizona had been largely sceptical as to the rigor of the empirical validation of the Freudian corpus to date. But I also defended psychoanalysis strenuously against the more damning charge, leveled by Karl Popper, of being altogether untestable and hence even unworthy of serious scientific consideration. Afterward, a certified psychoanalyst who is a senior professor of psychiatry there rose to rebut what she characterized as my "scurrilous attack" on Freud's theory. Referring to Freud's own account of the reasons *and* causes of opposition to

75

0028–7113/80/0039–0075 $1.75/2 © 1980, NYAS

psychoanalysis, she *overlooked* his explicit allowance for "those resistances to psycho-analysis that . . . are of the kind which habitually arise against most scientific innovations of any considerable importance."[1]

Instead of recognizing that Freud had thus made provision for the existence of rational, evidential motivations for scepticism, she focused entirely on the fact that he had also identified irrational, extraevidential or purely psychological inspirations of resistance to his theory. According to Freud,[1] the latter motivations

> are due to the fact that powerful human feelings are hurt by the subject-matter of the theory. Darwin's theory of descent met with the same fate, since it tore down the barrier that had been arrogantly set up between men and beasts. I drew attention to this analogy in an earlier paper (1917), in which I showed how the psycho-analytic view of the relation of the conscious ego to an overpowering unconscious was a severe blow to human self-love. I described this as the *psychological* blow to men's narcissism, and compared it with the *biological* blow delivered by the theory of descent and the earlier *cosmological* blow aimed at it by the discovery of Copernicus.

But as we saw, Freud himself had indeed made provision for the existence of rational, evidential motivations for scepticism toward psychoanalysis. Unfortunately at other times, he simply evaded the *cognitive* question of validation by pointing to the *de facto* growth of acclaim for his theories and/or his mode of psychiatric treatment. For example, he manifested just such an attitude when he commented on the challenge to *demonstrate* the therapeutic efficacy of analytic treatment: He was not only pessimistic regarding the feasibility of actually demonstrating the efficacy of psychoanalytic treatment but gave a strange twist to the cognitively unsolved question of efficacy by transmuting it into a sociological problem of resistance to new treatment modes, which will solve itself with the passage of time. Thus, Freud[2] wrote (in the unpolished English translation by J. Riviere):

> . . . the social atmosphere and degree of cultivation of the patient's immediate surroundings have considerable influence upon the prospects of the treatment.
>
> This is a gloomy outlook for the efficacy of psycho-analysis as a therapy, even if we may explain the overwhelming majority of our failures by taking into account these disturbing external factors! Friends of analysis have advised us to counterbalance a collection of failures by drawing up a statistical enumeration of our successes. I have not taken up this suggestion either. I brought forward the argument that statistics would be valueless if the units collated were not alike, and the cases which had been treated were in fact not equivalent in many respects. Further, the period of time that could be reviewed was too short for one to be able to judge of the permanence of the cures; and of many cases it would be impossible to give any account. They were persons who had kept both their illness and their treatment secret, and whose recovery in consequence had similarly to be kept secret. The strongest reason against it, however, lay in the recognition of the fact that in matters of therapy humanity is in the highest degree irrational, so that there is no prospect of influencing it by reasonable arguments. A novelty in therapeutics is either taken up with frenzied enthusiasm, as for instance when Koch first published his results with tuberculin; or else it is regarded with abysmal distrust, as happened for instance with Jenner's vaccination, actually a heaven-sent blessing, but one which still has its implacable opponents. A very evident prejudice against psycho-analysis made itself apparent. When one had cured a very difficult case one would hear: "That is no proof of anything; he would have got well of himself after all this time." And when a patient who had already gone through four cycles of depression and mania came to me in an interval after the melancholia and three weeks later again began to develop an attack of mania, all the members of the family, and also all the high medical authorities who were called in, were convinced that the fresh attack could be nothing but a consequence of the attempted analysis. Against prejudice one can do nothing as you can now see once more in the prejudices that each group of the nations at war has developed against the other. The most sensible thing to do is to wait and allow them to wear off with the passage of time. A day comes when the same people regard the same things in

quite a different light from what they did before; why they thought differently before remains a dark secret.

It is possible that the prejudice against the analytic therapy has already begun to relax.

As a response to the challenge that he provide genuinely cogent evidence for the therapeutic efficacy of analysis, Freud's statement here is deplorably question–begging and evasive. Furthermore, previously withheld assent to a theory, finally given for avowedly unknown reasons cannot be claimed to redound to the theory's *evidentially warranted* credibility. Even if such unexplained assent becomes widespread, it cannot cogently be held to count in favor of the theory, anymore than ill-founded initial prejudice can validly count against it. Nonetheless even ill-founded prejudices for *or* against a theory may be evidentially *fruitful* as follows.

An emotional revulsion felt for a theory, and *alternatively* an attraction toward it, may *each* be conducive to uncovering *evidentially* relevant information. Hostility toward the theory and/or toward its advocates may inspire a successful and useful search for *objectively contrary* evidence. But by the same token, the desire to embrace a theory and/or loyalty to its exponents may issue in ferreting out bona fide supporting evidence. Indeed, even the desire to *refute* a hypothesis may beget the unearthing of evidence objectively *favorable* to it after all. As I recall from a lecture given by the Yale scientist Hammond, he had conducted his pioneering research on the effects of cigarette smoking on humans in order to *refute* the conjecture that smoking is harmful: Having been a chain smoker for years, he hoped for evidence providing the comfortable assurance that he may continue to enjoy his four packs per day with impunity. But *malgré lui,* he then stumbled on stubborn evidence that drove him to the validation of a link between (heavy) cigarette smoking and lung cancer! Again the passionate desire to find evidence *favorable* to a given hypothesis in a certain domain of occurrences *may* serve to uncover highly disconcerting negative evidence.

So much for the possible heuristic benefits of emotional prejudice.

Now let us consider the merits of purely psychological appraisals of the motives for the rejection of theories like psychoanalysis. Unfortunately, many psychoanalysts still need to be reminded that it is illicit simply to *assume* the theory whose truth is first at issue, and then to invoke this very theory as a basis for a psychologistic dismissal of *evidential* criticism of its validity. Freud did not claim that psychoanalytic doctrine was handed to Moses on Mt. Sinai, and even if he had, the rest of us may be forgiven for nonetheless asking concerning the evidence for it. One wonders how those psychoanalysts who *dismiss* evidential criticism in the stated psychologistic fashion react to the following analogously question-begging *theological* argument: Even insufficient or *prima facie contrary* evidence for the existence of God does *not* impugn His existence, because God is merely *testing us* by giving us insufficient evidence of His presence! In this vein, Martin Buber invoked the doctrine of "the eclipse of God" to reconcile the horrors of the holocaust with the goodness of God amid attributing any goodness in the world to divine beneficence.

As recently as 1948, the psychoanalyst Robert Fliess—who is not to be mistaken for Freud's own confrère Wilhelm Fliess—gave an ominously totalitarian twist to his advocacy of *psychologistically dismissing* the doctrinal fractionation of the Freudian movement into splinter groups and even into avowedly rival schools of thought.[20] He points out that even a highly trained analyst—no less than a run-of-the mill analysand—may be neurotically resistant to the import of a Freudian tenet for his self-image. Especially when a future analyst is still undergoing his own training analysis, he may be prompted to question or even reject such a threatening Freudian tenet. Thus, the future analyst may:

vary the truth, instead of confirming it through improved introspection. In other words, his resistance may acquire the form of "dissension." He may, of course, do so in any phase of his education. He may, for instance, already working as an analyst, be confronted with equally unacceptable data by his patient, and be compelled by the consequent imminence of an empathic disturbance to gravitate, in the interest of his own psychic equilibrium, in the direction of an attenuated version of the theory of the unconscious. Or he may, in planning his training, unwittingly sidestep an as yet merely potential conflict of the same nature, by selecting a school of dissension, whose very existence derives from the same predicament in its founder. For it is here that the origin of dissension must be sought. The impact of psychic forces activated within an investigator in the course of his work may cause them to be reflected upon the collective object of his investigation and to direct his theoretical thinking. And if the personality of the individual thus imposed upon is a strong one, he is apt to break the ties of an inadequate education, and effect the foundation of a new school of psychological thought.

There are, naturally, concurring motives for establishment of these "schools." The historical closeness to Freud, the consequent "personalization" of learning, and the demand to accept a whole discipline practically from the hands of one man of genius, are a challenge to anyone's independence of mind. It is this intellectual independence which has indeed not infrequently led the dissenter-to-be, before he yielded to the public demand for the expurgation of psycho-analysis, to notable contribution to psycho-analysis in the sense in which it is here discussed. Yet it has never done so thereafter. For the denial of any one of the basic and interdependent facts found by Freud cannot but cause a defective crystalliza- tion of thought around the hollow nucleus of negation . . . .

If, at a future time, the extravagant growth of contemporary psychological teaching should be pruned back to the live stem of observation and theory of the first, the Freudian, period of its existence, psycho-analysis will regain its original independence of the precon- ceptions of the general as well as the learned public. It will then acquire the status of a scientific discipline comparable to others. Scholars from various fields will be given the opportunity to become competent in it, and, putting an old Jesuitic practice to secular purpose, will school themselves in both disciplines, psycho-analysis and their own. And an elaborate post-graduate education will be preventive of wasteful effort by including in its requirements that dissension be generally subjected to clearance in an analysis supplemen- tary to the training analysis of the dissenter.[20]

Invoking the Freudian doctrine of neurotic resistance to the recognition of buried inner conflicts, R. Fliess assumes in egregiously question-begging fashion that Freud always knows best, no matter what future evidence might be offered critically by an analyst at any stage of his professional development. Clearly, Fliess's automatic and *advance* discounting of *all* scepticism, however documented, as being solely inspired by neurotic resistance turns the validation of psychoanalysis into a logically circular *self*-validation. And it apparently did not occur to him—as Freud himself was driven to appreciate à propos of his seduction etiology of hysteria and of his sexual etiology of obsessional neurosis—that this vicious circle can be broken as follows: One can utilize the probative import, be it favorable or adverse, of evidence from the kinds of *extra*-clinical events that simply do not provide scope for the contaminating intrusion of resistance.

Alas, Fliess's rationale for his cowardly euphemistic "clearance" of dissenting intellectual independence is highly reminiscent methodologically of the candidly labeled "sacrifice of the intellect" enjoined by Ignatius of Loyola as the highest grade of obedience (in his 1553 Letter on Obedience to the Jesuits of Coimbra). And Fliess's proposed management of dissent is a cognate of the penchant of Soviet security organs to view political dissent as a psychiatric problem. On the other hand, no less authoritative a psychoanalytic spokesman than Ernest Jones repudiated the paternal- istic authoritarianism espoused so patronizingly by Fliess. Speaking of the conclusions reached by Freud on the basis of his investigations, Jones declared: " . . . it is plain that we should be forsaking the sphere of science for that of theology were we to regard these conclusions . . . as being sacrosanct and eternal."[21]

In the same vein, the well-known analyst Edward Glover, deploring precisely the dismissive appeal to "resistance" employed by Fliess, writes:[44]

> It is scarcely to be expected that a student who has spent some years under the artificial and sometimes hothouse conditions of a training analysis and whose professional career depends on overcoming 'resistance' to the satisfaction of his training analyst, can be in a favourable position to defend his scientific integrity against his analyst's theories and practice. And the longer he remains in training analysis, the less likely he is to do so. For according to his analyst the candidate's objections to interpretations rate as 'resistances'. In short there is a tendency inherent in the training situation to perpetuate error. Such a state of affairs clearly calls for the application of special safeguards.

Yet it is unclear not only just how the latter safeguards are to function but also what Glover expects from their employment: He implicitly provides ammunition for Fliess' dismissive stance by speaking of "our proven aetiological systems,"[45] despite having acknowledged that the analyst's interpretations of patient responses cannot be reliably checked and thus constitute "the Achilles heel" of psychoanalytical investigation.[46]

If one accepts Ernest Jones' own account[37] of the intellectual and personal rift between Freud and Adler, Freud's conduct toward Adler did not violate Jones' aforecited lofty methodological injunction. Speaking of the fact that Adler left the Vienna Psychoanalytic Society and formed his own splinter "Society for Free Psychoanalysis," Jones declares[38] that "the freedom of science . . . is certainly a worthy cause." But Jones adds:[38]

> The only issue was whether it was profitable to hold discussions in common when there was no agreement on the basic principles of the subject-matter; a flat-earther can hardly claim the *right* to be a member of the Royal Geographical Society and take up all its time in airing his opinions. Adler had drawn the correct inference by resigning. To accuse Freud of despotism and intolerance for what had happened has too obvious a motive behind it to be taken seriously.

Yet further documentation from the minutes of the Vienna Psychoanalytic Society shows the following:[39] While Freud did indeed offer a relevantly argued rebuttal to Adler's critique of his theory, Freud *also* engaged in a *question-begging* advance psychologistic dismissal of any assent to Adler's doctrine by others. Freud did so when he predicted that this dissident doctrine "will make a deep impression and will, at first, do great harm to psycho-analysis." Said he:[40] "It offers general psychology. It will, therefore, make use of the latent resistances that are still alive in every psychoanalyst, in order to make its influence felt."

According to Colby's account,[41] not only Freud's conduct but that of the membership of the Vienna Psychoanalytic Society was without blemish at least to the following extent: Far from expelling Adler from membership in the Society, as alleged by the biographers P. Bottome and F. Wittels, a majority of that Society voted to express regret over Adler's departure from it. But even the Freudian partisan Jones acknowledges[42] that, at a special plenary session of the Society on October 11, 1911 (Meeting #146) at which Freud announced the resignation of Adler and of three others, the following was voted by a majority of eleven to five: No one is to belong to both that Society and to the dissident one founded by Adler. And as Jones notes,[43] this affirmation of "a strong desire for a clean break" then immediately issued in the resignation of the remaining six pro-Adlerians from the original Society.

Having cited the version of the Freud-Adler rift furnished by the pro-Freudian authors Jones and Colby, it behooves us to summarize the salient points from the documented account just furnished by the historian of science Janet Terner and the psychiatrist W. L. Pew, who are Adlerian partisans.[27]

In 1910, Alfred Adler became president of the Vienna Psychoanalytic Society as well as coeditor (with Wilhelm Stekel) under Freud of a new journal for psychoanalysis. And early in 1911, at Freud's invitation, Adler presented his critique of Freud's sexual theory in a series of three lectures. After a mass denunciation of Adler by the Freudians that was reportedly "almost unequaled in its ferocity,"[28] Adler resigned his editorship as well as from the Vienna Psychoanalytic Society in the summer of 1911. And members of that Society who sided with Adler began informal gatherings with him at the Café Central. But at the next meeting of the Freudian group that fall

> Hanns Sachs read the indictment against Adler, and moved that it was incompatible to belong to both groups . . . the motion was carried and [the] six Adlerians rose, left, and went to the Café Central where [they] celebrated with Adler.[29]

Moreover

> Freud proscribed the quoting of Adler in any paper published by a Freudian (although Freud himself polemically railed against Adler whenever he chose). Indeed, one can examine the vast literature produced by the Freudians over the decades and rarely find Adler mentioned. Yet strangely, Adler's concepts often appear in thinly disguised form and in some cases almost verbatim renditions.[30]

Still worse,

> Freud never forgave Adler's dissidence, and in the years that followed he wielded his eloquent pen to encourage his loyal followers to deny or discredit Adler's discoveries—to keep Adler forever in the shadow of his own greatness. Freud knew the power of legend, and when he wrote his history of the psychoanalytic movement, which was reprinted in the United States [footnote omitted] in 1916, he presented his own prejudiced version of the split with Adler, scurrilously attacked him, and judged his ideas as "radically false." In this way, he got his account of these events into the history books long before it seemed important to anyone else [footnote omitted]. This affected Adler's image in America both immediately and in the long run.[31]

As an example of the hostile reception accorded to Adler by the power brokers of psychoanalysis in the United States, Terner and Pew[32] cite the following 1916 indictment, which presages Robert Fliess' aforecited *psychologistic* dismissal of all doctrinal apostasy by one-time Freudians:

> Dr. James J. Putnam, a venerable Boston Brahmin and champion of Freud, wrote: "A great longing has been felt by many conscientious students of human nature to find some way of escape from accepting Freud's conclusions . . . . To such persons Adler's mode of explanation is only too attractive. In plain terms, it offers a weapon with which Freud may be conveniently struck down by those . . . so minded [footnote omitted].

To illustrate just how "the 'hot' battle in Vienna was transplanted to America as a cold shoulder," Terner and Pew write:[33]

> The impact of Freud's curse on Adler was notable, and when Adler's *Study of Organ Inferiority* and *The Neurotic Constitution* appeared the following year, the reviews in the psychiatric literature were few and terse, with one notable exception. . . .
> But the die was essentially cast for Adler's place within American psychiatry. . . . To mention Adler or openly advocate his ideas was a risky stance against the mainstream of the profession.

No wonder, therefore, that when Rudolf Dreikurs—who was to become a vigorous opponent of the Freudian *monopoly* in American psychiatry—arrived in New York in 1937, "He was firmly warned not to declare himself an Adlerian,"[34] since *Freudian* psychoanalysis had become the mainstream in psychiatry and indeed had virtually achieved hegemony over the entire spectrum of the mental health professions:[35]

To be part of the mainstream was considered vital to most practicing psychiatrists. Its leaders were the power brokers who held the key to hospital appointments, professorships, and publishing opportunities—in sum—to recognition and success.

And by then, the psychologistic dismissal of opposition to Freud's ideas by his disciples had become inveterate:[36]

> As psychoanalysts became preeminent, they grew less tolerant of opposing points of view. Psychoanalysis was loudly acclaimed as a scientifically proven body of theory and practice, and those who actively challenged or opposed it within the professions were discredited as the simple unwashed—that is, the unanalyzed and therefore unknowing and superficial. Even electicism became ensnared in the defense of psychoanalysis. Since psychoanalysis was regarded as an established truth, it therefore became "uneclectic" and dogmatic *not* to acknowledge its basic truths. In other words, if these truths were acknowledged and the catechism of Freudian phrases was repeated, one could acquire the luster of revered "objectivity."

In any case, purely psychological, *extra*-evidential explanations of resistance to psychoanalysis would become relevant to understanding its rejection, *if* the presupposition of the following question *were* in fact true, which it is *not*: In the face of the doubting Thomases' explicit admission that there is indeed strong supporting evidence for the theory, why do they nevertheless still deny its credibility? By the same token, if Mr. X rejects atheism in favor of theism amid saying himself that the pertinent evidence does favor atheism, then we can try to understand his rejection in purely extra-evidential psychological terms. But before inquiring into purely psychological motivations for the rejection of either Freudism or any other "ism," the validity of the given theory must be adjudicated on the basis of the balance of the weight of the evidence. Until and unless this is done, the invocation of purely psychological, extra-evidential explanations for *either* the rejection *or* the acceptance of the theory runs the risk of begging the question of its validity, if only because *either attitude may well be prompted by relevant evidence*!

In saying this, I allow, of course, that the *available* evidence may warrant the rejection of an actually true theory, or alternatively, the acceptance of an actually false one.* But for the limited purpose of our present inquiry into understanding the rejection or acceptance of a theory vis-à-vis its validity in the light of the evidence, we can simple-mindedly lump together true beliefs with evidentially warranted ones on the one hand, and false ones with evidentially unwarranted ones on the other. Then I can say that psychological causation *as such* does not discriminate between valid beliefs and invalid ones. As I wrote elsewhere:[3]

> ... both true beliefs and false beliefs have psychological causes. The difference between a true or warranted belief [on the one hand] and a false or unwarranted one [on the other] must therefore be sought *not* in *whether* the belief in question is caused; instead, the difference must be sought in the particular *character* of the psychological causal factors which issued in the entertaining of the belief; a *warrantedly held belief, which has the presumption of being true, is one to which a person gave assent in response to awareness of supporting evidence.* Assent in the face of awareness of a *lack* of supporting evidence is irrational, although there are indeed psychological causes in such cases [as well] for giving assent. Thus, one person may be prompted to give assent to a certain belief solely because this belief is wish-fulfilling for him, while another may accept the same conclusion in response to his recognition of the existence of strong supporting evidence. And the belief *may* be true.

---

*That is why there are what physicians call respectively "false negatives" and "false positives," errors to which statisticians refer as "Type I" and "Type II" errors respectively.

More generally, as I wrote in the same essay,[22]

the causal generation of a belief does not, of itself, detract in the least from its truth. My belief that I address a class at certain times derives from the fact that the presence of students in their seats is causally inducing certain images on the retinas of my eyes at those times, and that these images, in turn, then cause me to infer that corresponding people are actually present before me. The reason why I do not suppose that I am witnessing a performance of *Aïda* at those times is that the images which Aïda, Radames, and Amneris would produce are not then in my visual field. The causal generation of a belief in no way detracts from its veridicality. In fact, if a given belief were not produced in us by definite causes, we should have no reason to accept that belief as a correct description of the world, rather than some other belief arbitrarily selected. Far from making knowledge either adventitious or impossible, the deterministic theory about the origin of our beliefs alone provides the basis for thinking that our judgments of the world are or may be true. Knowing and judging are indeed causal processes in which the facts we judge are determining elements along with the cerebral mechanism employed in their interpretation.

On this causal conception of the generation of (perceptual) beliefs, there are typically at least some *extra*-ideational causes of the occurrence of the following kind of awareness-state: A mental state that is at once the evidential source (or "reason") *and* the psychological cause of entertaining (or espousing) a belief. Thus, the "*initial*" causal promptings of our beliefs concerning the external world are typically extra-ideational events rather than mental apprehensions of evidential reasons. Hence I must reject the upshot of the following account offered by C. S. Lewis:[23]

All beliefs have causes but a distinction must be drawn between (1) ordinary causes and (2) a special kind of cause called 'a reason'. Causes are mindless events which can produce other results than belief. . . . A belief which can be accounted for entirely in terms of causes is worthless. This principle must not be abandoned when we consider the beliefs which are the basis of others. Our knowledge depends on our certainty about axioms and inferences. If these are the result of causes, then there is no possibility of knowledge. Either we can know nothing *or* thought has reasons only, and no causes.
    . . . All attempts to treat thought as a natural event involve the fallacy of excluding the thought of the man making the attempt.
    It is admitted that the mind is affected by physical events . . . But thought has no father but thought. It is conditioned, yes, not caused . . . .
    The same argument applies to our values, which are affected by social factors, but if they are caused by them we cannot know that they are right.

On this basis, Lewis[24] feels entitled to give an affirmative answer to his question "Does 'I know' involve that God exists?". He invokes God as the very source rather than merely as the epistemological underwriter of our knowledge. Whereas Descartes' epistemological specter was the evil deceiving genius, Lewis' corresponding nightmare is constituted by "mindless events": Allegedly such events cannot causally induce bona fide knowledge states in us because "thought has reasons only, and no causes." How then can he hope to show that the inclusion of initially "mindless" causation does not significantly contribute to the *warrant* for the following well-taken assertion by him?: "Suppose I think, after doing my accounts, that I have a large balance at the bank. And suppose you want to find out whether this belief of mine is 'wishful thinking'. You can never come to any conclusion by examining my psychological condition."[25] But if so, is the initially "mindless" causation by corresponding actual bank assets held in his name not considerably more justificatory here than the much more inscrutable "Supernatural" of which he speaks?

But there is full agreement with C. S. Lewis[26] that, regardless of the motivation for holding a certain belief or disbelief in a given case, its *validity* must be assessed on the basis of the pertinent evidence, not by reference to the psychological motivation for entertaining it: It would be altogether fallacious, for example, to cast aspersions on *disbelief* in personal immortality on the purported psychological ground that this disbelief is caused by the "death instinct"—Freud's "thanatos." Such psychologistic dismissal of the rejection of personal immortality might invoke man's putative craving for death as a release from the sorrows of life. But even if all men could be shown to harbor such a "death instinct," this psychological fact could not itself preclude that (a) there might *also* be strong objective evidence for *disbelieving* in the *post mortem* existence of the self, and (b) even psychologically, actual disbelief in an after-life is explained by awareness of this evidence rather than by the putative death instinct.

Similarly, the proponent of psychoanalysis in Arizona patently *begged the question* of its validity, when she sought to *dismiss evidential criticism* on the basis of the following purported psychological explanation: Those who question the evidential credentials of Freudian theory do so because it poses a so-called "narcissistic" threat to man's ego by asserting the sovereign dominance of unconscious forces, at least over unanalyzed people. As an intended premise of an *argument* against disbelief in Freudian theory, this *dismissive psychological explanation* is thus clearly *irrelevant*. But furthermore, even as a purely causal account of resistance of psychoanalysis, this psychological explanation fails *empirically,* since it does *not simultaneously* accommodate the impressive widespread *acceptance* of Freudianism during the past quarter century, at least in the United States: After all, not only in influential literary circles but even in sizable segments of the educated lay public, psychoanalysis has become a cultural idol by commanding a kind of quasi-religious veneration.[4]

For example, in November 1977, the columnist Jack Anderson published a report of an hour's interview of President Carter entitled "What is Jimmy Carter Really Like?"[5] Anderson relates that after the interview, he submitted the transcript to the psychoanalyst Saretsky. And Anderson reports the conclusions reached by Dr. Saretsky after several days of study. One such product of Saretsky's expertise is that Carter is a man who believes in God *and isn't afraid to say so.* Presumably, if Anderson felt less deferent toward psychoanalysts, he would hardly have deemed it remarkable that a President of the United States who does believe in God *isn't* afraid to say so. Oddly enough there was no mention of Carter's earlier *Playboy* interview. Jack Anderson's recitation of a series of other *at best* trite and altogether safe comments from his analytic consultant typifies the halo that is often uncritically bestowed on psychoanalysis in our culture. Hence I am prompted to ask: If the "narcisssistic" threat to man's ego is held to account adequately for the vigor of such opposition to psychoanalysis as is encountered in some quarters, then how could Freud's ideas have so triumphantly swept aside these alleged defensive reactions in the culture at large, even among many of the unanalyzed? Thus it would seem that the all too facile invocation of ego-threat to explain *opposition* to psychoanalysis also boomerangs *empirically* by running afoul of the rather prevalent espousal of Freudianism.

In any case, the fallacy of *psychologistically* discrediting the *rejection* of Freudianism is on a par with *each* of the following two similarly dismissive gambits: (i) Endeavoring to undermine atheism by claiming that Madeline Murray–O'Hare and other atheists simply *hate* the idea of a personal God, and (ii) seeking to discredit theism by merely pointing out that its adherents derive much emotional solace from their belief in a cosmic protective father figure. Freud[6] commendably recognized in his book *The Future of an Illusion* that theism cannot validly be discredited by the latter dismissive gambit, and Erich Fromm usefully reiterated this point[7] in his book *Psychoanalysis and Religion.* We see that the attempt to undermine the theist's *acceptance* of belief in God purely *psychologistically* because it is wish-fulfilling is just as ill-conceived logically as the correspondingly dismissive endeavor to cast psychologistic aspersions on the atheist's *rejection* of that belief. Surely there are some beliefs that are both true and wish-fulfilling for at least a good many people, no less than there are other true beliefs that are wish-contravening.

Mutatis mutandis, efforts to discredit the *espousal* of Freudianism by facile psychologistic devices are no less unsound logically than appealing to the ego-threat hypothesis as a basis for impugning intellectual *opposition* to psychoanalysis. I am concerned to stress this *equi*-fallaciousness. But so far, I have been at pains one-sidedly to indict the use of mere psychologism for dismissing the *rejection* of Freudian theory. Hence I now wish to call attention to the like unsoundness of the following peremptory psychologistic dismissal of the *acceptance* of psychoanalysis: Freud's bizarre adult psychoanalytic beliefs can be discounted without regard for his suppor-tive arguments because—as we know from his biography—he experienced the traumatic childhood event of having entered his parents' bedroom while they were locked in a sexual embrace and of having been irately ordered out by his father.[8] By the same token, it will not do at all to try to impugn the therapeutic efficacy of Freudian treatment by offering the following *motivational* explanation of why anyone chooses to become a psychoanalyst: Analysts are driven by a kind of generalized voyeurism, coupled with morbid curiosity or are just socially-sanctioned peeping Toms. Unfortunately, none other than Thomas Szasz saw fit to stoop to the nadir of the psychologistic dismissal of Freud's theories: In the manner of gutter journalism, he depicts these theories as mere shams, purportedly owing their inspiration to Freud's desire to be "The Jewish Avenger" vis-à-vis the gentile world.[47]

Recently the philosopher Frank Cioffi[9] has offered a *differently-argued* yet motivational critique of Freud's championship of psychoanalysis. What makes Cioffi's critique interesting is that though motivational, it is *not peremptorily* psychologistic. Thus Cioffi tried to *show* that while Freud seemingly went through the motions of being engaged in an explanatory inquiry, the Viennese doctor's ostensible arguments become *intelligible* only after the following is recognized: His hypotheses are prompted not by a concern with logically relevant evidence, but by his idiosyncrati-cally intense personal preoccupation, presumably with sexuality.

But does Cioffi give cogent grounds for his claim that Freud's reasoning is unintelligible, unless we *exclude* concern with pertinent evidence from Freud's actual motivation for espousing psychoanalysis? I shall briefly scrutinize just a couple of the considerations adduced by Cioffi for this claim in order to show why I deem it to be ill-founded. Thus I shall reject Cioffi's grounds for resorting to a purely psychological explanation of Freud's own advocacy of psychoanalysis. Cioffi sees himself as having resorted to explaining Freud's rationale in terms of *extra*-evidential motives only *after* having ruled out logically pertinent evidential promptings. But I shall now illustrate how he mishandled his examination of Freud's reasoning and was thereby driven to the gratuitous or mistaken conclusion that concern with pertinent evidence had played no essential role in Freud's rationale for espousing psychoanalysis.

Cioffi[10] gives the following lucid statement of his thesis:

> Freud behaves neither like someone who is addressing himself to the problem of the causes and nature of the neuroses but bungles the job from incompetence or lack of methodological sophistication, nor like someone who is stymied by the intrinsic difficulties of the problem, but rather like someone who, while going through the motions of engaging in an explanatory enquiry, reveals in an enormous variety of ways that he has other ends in view. (Fine shades of misbehaviour.)
>
> One often comes across people whose preoccupations with a putatively explanatory factor is ostensibly derived from their interest in its pathogenic potentialities, but is really intrinsic. The majority of those who speculate as to whether slums, or the decline in church-going are causes of delinquency are not really interested in delinquency, but are

interested in slums and the decline in church-going. What is noteworthy in Freud is the way in which the prestige of aetiological and prophylactic enquiries is exploited in the interest of an idiosyncratic preoccupation (or, perhaps I should say, an idiosyncratically intense pre-occupation).

Having adduced Freud's psychobiography of Leonardo da Vinci in support of this thesis, Cioffi[11] writes as follows:

> As another instance of the ease with which the variety of mechanisms at Freud's disposal enables him to press into the service of the thesis of infantile pathogenicity, whatever parental circumstances the childhood history of his subject happens to provide, consider his account of how inevitable it was, given the character of Dostoyevsky's father, that he [the son] should have come to possess an over-strict super-ego:
>> If the father was hard, violent and cruel, the super-ego takes over these attributes from him, and in the relations between the ego and it, the passivity which was supposed to have been repressed is re-established. The super-ego has become sadistic, and the ego becomes masochistic, that is to say, at bottom passive in a feminine way. A great need for punishment develops in the ego, which in part offers itself as a victim to fate, and in part finds satisfaction in ill-treatment by the super-ego (that is, the sense of guilt).

This is not at all implausible. But neither is this:
> The unduly lenient and indulgent father fosters the development of an over-strict super-ego because, in the face of the love which is showered on it, the child has no other way of disposing of its aggressiveness than to turn it inwards. In neglected children who grow up without any love the tension between ego and super-ego is lacking, their aggressions can be directed externally . . . a strict conscience arises from the co-operation of two factors in the environment: the deprivation of instinctual gratification which evokes the child's aggressiveness, and the love it receives which turns this aggressiveness inwards, where it is taken over by the super-ego.

As I see it, the conjunction of these two quotations from Freud is tantamount to the following assertion of *causal sufficiency*: If a child has a father who is either "hard, violent and cruel" or "unduly lenient and indulgent," then it develops a hyper-strict super-ego. Yet immediately after having given us the two Freud quotations in question, Cioffi draws the following astonishing inference:[12]

> That is, if a child develops a sadistic super-ego, either he had a harsh and punitive father or he had not. But this is just what we might expect to find if there were no relation between his father's character and the harshness of his super-ego.

But Cioffi employs a gross logical sleight-of-hand when he trivializes Freud's stated *bicausal* claim into one that does *not* affirm any causal relevance of the father's character to the child's super-ego. For Cioffi replaced the nontrivial antecedent that the child's father was either harsh or *unduly lenient*—which is a *non*exhaustive disjunction—by the utterly trivial consequent that either the child had a harsh father or he had not. And the latter disjunction is, of course, trivial because it is exhaustive besides being mutually exclusive. In other words, Cioffi has speciously replaced the paternal property of undue leniency, which Freud had invoked causally, by the logically much weaker property of merely being *non*harsh. And amid committing this sleight-of-hand, he likewise overlooked the following: Freud's bicausal assertion had affirmed the causal *sufficiency* of the specified paternal traits for hyper-strict childhood super-ego development, but *not* their being causally *necessary* for such super-ego development. For Cioffi gratuitously depicts Freud as having legitimated a retrodictive causal inference *from* the child's super-ego structure *to* that of the father, whereas the Freudian quotations adduced by Cioffi countenance only an inference in the opposite direction.

By parity with Cioffi's fallacious reasoning one could deduce the absurdity that getting shot is causally irrelevant to being killed from the following sound bicausal assertion: If a person is either shot in a vital organ or massively poisoned by a fast-acting toxin *without* being shot in a vital organ, then he will be killed in all likelihood. By parity of reasoning with Cioffi's argument, this would become: If a person is killed, then he was either shot or he wasn't, which is just what we would expect, if there were no relation between getting shot and getting killed.

As a second brief illustration of Cioffi's attempt to document Freud's alleged indifference to the actual evidence, let me cite another argument by him. Cioffi writes:[13]

> It might seem that there can be no question of the genuinely empirical-historical character of those clinical reconstructions which incorporate references to the external circumstances of the patient's infantile life, such as that he had been threatened with castration or been seduced, or seen his parents engaged in intercourse. These at least are straightforwardly testable, and their accuracy would therefore afford evidence of the validity of psychoanalytic method; for if the investigation into the infantile history of the patient revealed that he had had no opportunity of witnessing intercourse between his parents (the primal scene), or that he [had] not been sexually abused, or not threatened with castration, this would cast doubt on the validity of the interpretative principles employed and on the dependability of the anamnesis which endorsed them.
>
> But Freud occasionally manifests a peculiar attitude towards independent investigation of his reconstruction of the patient's infantile years. In 'From the history of an infantile neurosis' he writes: 'It may be tempting to take the easy course of filling up the gaps in a patient's memory by making enquiries from the older members of the family: but I cannot advise too strongly against such a technique . . . One invariably regrets having made oneself dependent on such information. At the same time confidence in the analysis is shaken and a court of appeal is set up over it. Whatever can be remembered at all will anyhow come to light in the course of further analysis.'

But does Freud here reject all use of independent external evidence to test the historical veracity of his clinical reconstructions of a patient's infancy *after* the completion of the psychoanalysis? It would seem not. What Freud does renounce here is questioning senior relatives, conducted by the analyst to fill up the gaps in the patient's memory as a procedural technique for making progress in the analysis. This renunciation does not preclude his willingness to test his clinical reconstructions by means of independent external evidence once the analysis has been completed. Indeed, he displayed this kind of willingness in his retrospective external evaluation of his clinical findings in his "Rat Man" case. For in this case, he was prompted by just such independent historical evidence to abandon his prior hypothesis as to the specifics of the sexual aetiology of adult obsessional neurosis. Corresponding remarks apply to his abandonment of his seduction etiology of hysteria, as we know from his Letter 69 to Wilhelm Fliess, dated September 21, 1897.

Indeed this reading of Freud is strengthened upon taking account of the sentences that Cioffi omitted tendentiously from his quotation. The latter is taken from a footnote that Freud appended [cf. Collected Papers, tr. A. and J. Strachey, 1959, Basic Books, New York, vol. 3, pp. 481–2, fn. 2] to the statement that in the Wolf Man's later years, this patient "was told many stories about his childhood." Freud's appended footnote *begins* with the sentence "Information of this kind may, as a rule, be employed as absolutely authentic material," a sentence that immediately precedes the one with which Cioffi begins his quotation! Moreover, the sentence whose omission from *within* his quotation Cioffi does indicate reads as follows: "Any stories that may be told by relatives in reply to inquiries and requests [from the analyst] are at the mercy of every critical misgiving that can come into play." The latter omission supplies Freud's *epistemological reason* for *not* succumbing to "the easy course of filling up the gaps in a patient's memory by making inquiries from the older members of his family": Whereas, "as a rule," stories that the patient's relatives told the patient *spontaneously* in his later years are "absolutely authentic," responses by relatives to pointed inquiries from the analyst may well be quite contaminated by misgivings.

Hence, I believe that Cioffi's motivational critique of Freud misfires by being ill-founded. But this is *not* to say that *psychological* explanations of the acceptance of psychoanalysis are *always* misdirected.

For suppose it *were* now agreed that there is in fact a serious dearth of objective evidential support for psychoanalysis. And assume further that some educated people who do acknowledge this sparsity of validating evidence nonetheless vigorously espouse Freud's theory. Of course in that case we cannot attribute their assent to their awareness of cogent supportive findings, or to their belief that they possess such favorable evidence. Hence it then becomes well-nigh imperative that we ask: In the face of the avowed scarcity of such validating findings, why does anybody nonetheless embrace the theory? And plainly, under these conditions, it is hardly question-begging or misdirected to ask the following advisedly loaded question: What nonevidential psychological motivations prompt otherwise rational people to believe the fable brilliantly concocted by the genius whom Nabokov called "the Viennese witch doctor?"

Quite naturally, therefore, sceptics who do decry Freudian theory as a fable or pernicious myth have asked just this question. And some of them have suggested some possible answers. Let me conclude by concisely outlining four of their proposed explanations, although I do not venture to guess at their relative potential importance.

(a) Psychoanalysis as a general theory of man and even liberating gospel became the secular equivalent or alternative of traditional religion for many intellectuals who had abandoned theism. Being such a substitute religion, it commanded the fierce loyalty often displayed by those who are converted to a new religion during adulthood.

(b) As Salter[14] has pointed out:

> Writers as a group are probably among the most neurotic in the population. When word of the new panacea for their troubles with old wives or new books drifted through they went for the treatment in a big way. Many of them were encouraged to write about their analysis as a means of paying for the expensive treatment. A rash of novels, plays, and short stories resulted. Some even wrote lengthy magazine articles embedded with a remarkable variety of rather too-candid clinical detail.
>
> The net result was a public relations campaign that millions of dollars could not have duplicated. Once analysis became fashionable among the writers, it was a brief step before their more impressionable readers were fretting impatiently in the analysts' busy waiting rooms.

(c) The favorable retrospective valuation of the personal analyses undergone by many professional people is often assured by the expense, time and emotional pain of an analysis. "There is abundant evidence that where we have made a sacrifice to obtain some object, we come to value it: we cannot afford to admit to ourselves that we have made the sacrifice in vain."[15] It is small wonder that relatively few people are willing to admit to themselves that they may have made a costly mistake in the sense that their actual therapeutic gains may hardly be commensurate with the overall cost. *A fortiori,* this mechanism is operative to allay such doubts as may be developed by analysts themselves, who have not only undergone even more arduous training analyses but have the vested interest of a life-time career commitment and livelihood in the practice of psychoanalysis. And even if analysts waver nonetheless—as a minority of them do from time to time— there are other influences at work to quell or mitigate their misgivings.[16] Not least of these is the intoxication from their socially sanctioned prestige as presumed medical healers of the mind, which is also reflected by the dependence evinced toward them by their patients.

(d) Freud, who began to publish his ideas towards the end of the Victorian age took the obscenity out of sex and helped many to jettison some of the oppressive guilt they had felt about their sexual urges.[17] Moreover, Freud was an incredibly brilliant, dramatic, captivatingly imaginative and disarmingly persuasive expositor who rested his case not only on his clinical findings but also on keen perceptions of daily life which sometimes ring particularly true. Moreover, his wide learning enabled him to draw on anthropology, literature, and religion, and even on his analyses of the lives of great cultural figures. As Sutherland[18] put it concisely: "He tries in fact to make sense in his own terms of the whole of human existence, and there is a temptation to be awestruck by the size of the edifice at the expense of failing to notice whether it is built of bricks or cardboard."

But for the reasons I have given, such motivational considerations surely do not determine the actual scientific merits of Freudian psychogenics or therapeutics. Indeed, it is conceivable that some day there just *might* be objectively strong supportive evidence for Freudianism. Yet even then, most of its fervent adherents may actually be believing in it for nonevidential emotional reasons. And in that event, it clearly would yield a correct and enlightening psychological explanation of their assent to adduce these emotional factors. But, although this psychological explanation of belief would be illuminating, in the face of the posited evidential support this explanation could not also justify a psychologistic dismissal of Freudism.†

†For a discussion of the present-day scientific merits of Freudian psychoanalysis, see A. Grünbaum.[19]

ACKNOWLEDGMENTS

The author is greatly indebted to the Fritz Thyssen Stiftung for the support of research. Grateful acknowledgment is also made to the Center for the Humanities at the University of Southern California for permission to publish here an *enlarged* version of the paper by the same title from the Fall 1978 issue of *Humanities in Society*.

REFERENCES

1. FREUD, S. 1925. The resistances to psychoanalysis. *In: Collected Papers.* J. Strachey, Ed. **5** (1959): 173. Basic Books. New York, N.Y.
2. FREUD, S. 1949. *Introductory Lectures on Psychoanalysis* (first published in English in 1922). J. Riviere, Tr.: 386-387. Allen & Unwin. London, England.
3. GRÜNBAUM, A. 1972. Free will and laws of human behavior. *In: New Readings in Philosophical Analysis.* H. Feigl, K. Lehrer and W. Sellars, Eds.: 618. Appleton-Century-Crofts. New York, N.Y.
4. FISHER, S. & R. P. GREENBERG. 1977. *The Scientific Credibility of Freud's Theories and Therapy* : viii. Basic Books. New York, N.Y.
5. ANDERSON, J. 1977. What is Jimmy Carter really like? *Parade,* The Pittsburgh Press (Nov. 13): 9-11. Pittsburgh, Pa.

6. FREUD, S. 1927. The future of an illusion. *In: Standard Edition of the Complete Psychological Works of Sigmund Freud.* J. Strachey, Ed. and Tr. **21** (1961): 5-56. Hogarth Press. London, England.
7. FROMM, E. 1950. *Psychoanalysis and Religion*: 12n. Yale University Press. New Haven, Conn.
8. KIERNAN, T. 1974. *Shrinks, etc.* : 23. Dial Press. New York, N.Y.
9. CIOFFI, F. 1970. Freud and the idea of a pseudo-science. *In: Explanation in the Behavioural Sciences.* F. Cioffi and R. Borger, Eds. : 471-499, 508-515. Cambridge University Press. Cambridge, England.
10. CIOFFI, F. Ref. 9: 515.
11. CIOFFI, F. Ref. 9: 484-485.
12. CIOFFI, F. Ref. 9: 485.
13. CIOFFI, F. Ref. 9: 480.
14. SALTER, A. 1952. *The Case Against Psychoanalysis*: 11-12. Henry Holt. New York N.Y.
15. SUTHERLAND, S. 1976. *Breakdown*: 111. Weidenfeld & Nicolson. London, England.
16. SUTHERLAND, S. Ref. 15: 109-113.
17. SALTER, A. Ref. 14: 12-13.
18. SUTHERLAND, S. Ref. 15: 116.
19. GRÜNBAUM, A. 1977. How scientific is psychoanalysis? *In: Science and Psychotherapy.* R. Stern, L. Horowitz and J. Lynes, Eds. : 219-254. Haven Press. New York, N.Y. Also, GRÜNBAUM, A. 1979. Is Freudian Psychoanalytic Theory Pseudo-Scientific by Karl Popper's Criterion of Demarcation? *American Philosophical Quarterly.* **16**: 131.
20. FLIESS, R. 1948. Foreword. *In: The Psycho-analytic Reader.* R. Fliess, Ed.:XV-XVII. International Universities Press. New York, N.Y. I am indebted to Morris Eagle for this reference.
21. JONES, E. 1946. A valedictory address. *International Journal of Psychoanalysis.* **27**: 11.
22. GRÜNBAUM, A. Ref. 3: 617-618.
23. LEWIS, C. S. 1970. *God in the Dock.* Walter Hooper, Ed. : 275. William B. Eerdmans Publishing Co. Grand Rapids, Michigan. I am indebted to Rosamond Sprague for this reference.
24. LEWIS, C. S. Ref. 23: 274-277.
25. LEWIS, C. S. Ref. 23: 272.
26. LEWIS, C. S. Ref. 23: 272-273.
27. TERNER, J. & W. L. PEW. 1978. *The Courage To Be Imperfect, The Life and Work of Rudolf Dreikurs*: 38-39, 118-119. Hawthorn Books, Inc. New York, N.Y. I am indebted to Edward J. Shoben, Jr. for calling my attention to this work.
28. TERNER, J. & W. L. PEW. Ref. 27: 39.
29. TERNER, J. & W. L. PEW. Ref. 27: 39. This quotation is *part* of a citation for which they give the *Journal of Individual Psychology* **20** (1964): 124 as their source (on page 46, fn. 30).
30. TERNER, J. & W. L. PEW. Ref. 27: 39.
31. TERNER, J. & W. L. PEW. Ref. 27: 118.
32. TERNER, J. & W. L. PEW. Ref. 27: 118-119.
33. TERNER, J. & W. L. PEW. Ref. 27: 119.
34. TERNER, J. & W. L. PEW. Ref. 27: 125.
35. TERNER, J. & W. L. PEW. Ref. 27: 118.
36. TERNER, J. & W. L. PEW. Ref. 27: 124.
37. JONES, E. 1955. *The Life and Work of Sigmund Freud.* Vol 2: 129-134. Basic Books. New York, N.Y. I am indebted to Dr. Edward J. Shoben, Jr. for this reference as well as for the reference in item 39.
38. JONES, E. Ref. 37: 133.
39. NUNBERG, H. & E. FEDERN, Eds. 1974. *Minutes of the Vienna Psychoanalytic Society.* Vol. III: Scientific Meetings No. 125 (Jan. 4, 1911), No. 129 (Feb. 1, 1911), and No. 146 (Oct. 11, 1911). International Universities Press. New York, N.Y.
40. NUNBERG, H. & E. FEDERN, Eds. Ref. 39: 147. (Meeting No. 129, Feb. 1, 1911).
41. COLBY, K. M. 1951. On the disagreement between Freud and Adler. *The American Imago.* **8**: 237.

42. JONES, E. Ref. 37: 133–134.
43. JONES, E. Ref. 37: 134.
44. GLOVER, E. 1952. Research methods in psychoanalysis. *International Journal of Psychoanalysis* **8:** 403.
45. GLOVER, E. Ref. 44: 408.
46. GLOVER, E. Ref. 44: 405.
47. SZASZ, T. 1978. *The Myth of Psychotherapy.* Ch.7–9. Anchor Press. Garden City, N.Y.

AN ANALYSIS OF PSYCHOLOGICAL AND SOCIOLOGICAL AMBIVALENCE: NONADHERENCE TO COURSES OF ACTION PRESCRIBED BY HEALTH-CARE PROFESSIONALS*

Irving L. Janis

Department of Psychology
Yale University
New Haven, Connecticut 06520

This paper discusses the theoretical concepts and findings that have emerged from a research project on which I have been working for the past 10 years. The research program is designed to test hypotheses concerning the conditions under which a professional counselor will and will not have a positive influence on clients' adherence to a recommended course of action. Most of the studies are field experiments conducted in health-care settings. The recommendations typically involve short-term losses in order to attain long-term gains, such as giving up smoking, dieting, or undergoing a surgical operation. The objective of the research is to increase our understanding of when, how, and why the social influence of professionals on clients achieve their goals. Specifically, the studies are intended to enable us to explain and predict when a professional counselor will succeed or fail to help clients to adhere to a recommended course of action that both agree would be in the client's best interests.

My theoretical analysis for developing testable hypotheses took as its point of departure the concept of *reference power* as a major determinant of a person's social influence. This concept owes much to the work of Robert Merton (1957) on reference groups and reference individuals.

When Merton's book on *Sociological Ambivalence* was published in 1976, I discovered that another of his key concepts was applicable to my research on adherence and nonadherence. Most of my previous work dealt with psychological ambivalence. It focused on how variations in the counselor's behavior affected the clients' emotions, expectations, attitudes, and actions. But I soon realized that I was encountering sources of ambivalence that are "built into the structure of social statuses and roles," which Merton (1976, p.5) designates as *sociological* ambivalence.

When people are ambivalent, as Merton notes, they are "pulled in psychologically opposed directions," as when a person feels both love and hate for someone or is inclined both to accept and reject another's demands (1976, p.6). Latent ambivalence occurs, to some degree, in all our human contacts. When ambivalence becomes sufficiently intense to be manifest in our dealings with a particular person, our relationship with that person tends to become seriously impaired, if not terminated altogether.

Merton (1976, pp. 19-31 and pp. 65-72) gives a vivid account of sentiments unintentionally evoked by physicians who, in keeping with the requirements of their professional role, are more detached, more demanding, and more frustrating than their patients are willing to tolerate. He calls attention to various dysfunctional consequences that result from health-care professionals living up to their roles. These

*The research reported in this paper was supported by a research grant from the National Science Foundation.

0028–7113/80/0039–0091 $1.75/2 © 1980, NYAS

include patients' suspicion that they are being exploited financially, failures to see a physician when serious medical problems occur, hostile attitudes toward the entire profession, and disproportionate tendencies to initiate unwarranted malpractice suits. Incompatible role expectations also have corruptive influences on some of the troubled practitioners themselves who try to avoid generating ambivalence by "departing from what they know to be the most appropriate kind of medical care" (1976, p.72). Another dysfunctional consequence would be patients' nonadherence to prescriptions and health practices recommended by health-care practitioners.

I am now prompted to reconceptualize my theoretical analysis of practitioner-client relationships in terms of the linkage between the two types of ambivalence, taking account of Merton's plausible assumption that "sociological ambivalence is one major source of psychological ambivalence" (1976, p.7). The final section of this paper represents a first attempt at a reformulation in terms of a double focus on sociological and psychological ambivalence, as recommended by Merton. The need for linking the two sources of ambivalence will become more apparent after I discuss the main phenomena of non-adherence.

THE PROBLEM OF NONADHERENCE

Numerous studies indicate that patients often disregard physicians' recommendations for dealing with serious health problems. In a study of 47 men and women treated at an out-patient clinic in Liverpool, England, more than half of the medical instructions given by the physicians could not be recalled accurately by the patients immediately after they left the consulting room (Ley and Spelman, 1965). Investigations of American women with acutely ill children suffering from rheumatic fever, streptococcal pharyngitis, or otitis media found that from 34 to 82 percent were seriously endangering the health of their children by not giving them the proper doses of penicillin that had been explicitly prescribed by their physicians (Bergman and Werner, 1963; Chamey et al., 1967; Feinstein, et al., 1959). A study of 154 new adult patients in the general medical clinic of a large teaching hospital found that 37 percent failed to comply substantially with physicians' recommendations and only 14 percent complied fully (Davis, 1971). Reviews of the large number of studies on patients' failure to comply with physicians' recommendations report wide variation in different circumstances, with noncompliance rates ranging from 15 to 93 percent (Davis, 1966; Sacket and Haynes, 1977).

Medical practitioners sometimes express dissatisfaction with their professional life, despite the high prestige and deference accorded them, because a sizeable percentage of their patients fail to follow the regimens they prescribe (see Kasl, 1975). Many physicians avoid the disappointments of nonadherence by not treating patients with certain types of disorder, like hypertension, which require unpleasant medication as well as restrictions on smoking, drinking, and eating (see Maddox, Anderson, and Bogdonoff, 1966; Stamler, Schoenberger, Lindberg, et al., 1969).

Problems of nonadherence have also been well documented in studies of many clinics that offer professional help to heavy smokers and to overweight men and women. In the United States this is a burgeoning industry, quite profitable to private entrepreneurs. Each year hundreds of thousands of new consumers and recidivists pay for their services. But most of their successes appear to be short-lived. A high percentage of those who come to the clinics do cut down on cigarette smoking or lose weight for a few weeks, but most of them fail to adhere to the prescribed regimen after supportive contact terminates (see Atthowe, 1973; Sackett and Haynes, 1977; Shewchuck, 1976). In a review of the literature on the effectiveness of programs for heroin addicts and heavy drinkers, Hunt and Matarazzo (1973) found that, just as with heavy smokers, many people begin to abstain in response to whatever professional treatment they receive, but a very high percentage relapse within a month or two after starting the program.

Why do many people break off before completing the prescribed regimen or backslide after having changed their behavior temporarily? Why do other people successfully adhere to recommended treatments? Some of the pertinent factors have been tentatively identified. One obvious factor in backsliding is intensity of suffering or deprivation, which may be especially hard to take over a long period of time. But social and psychological factors also affect a person's tolerance for pain, frustration, and resisting temptations. Many studies have found that nonadherence is linked with difficulties in patients' relationships with health-care professionals. Disappointment of patients with their physicians has been found to lead to their not keeping subsequent appointments or breaking off the treatment entirely (Davis 1967; Vincent, 1971; Zola, 1973). These disappointment reactions are partly attributable to the physicians' behavior. Physicians often fail to find out the patients' reasons for wanting to be examined and fail to explain the purpose of the treatments they prescribe. These two kinds of deficiency have been singled out as specific sources of patients' failure to act in accordance with medical recommendations.

In a review of the extensive literature on adherence to recommended medical regimens among outpatients, Kasl (1975) concludes that a crucial source of difficulty resides in the conflicting expectations of the health-care professional and the patient, which involve divergent role expectations. In other words, studies of the psychology of adherence point to precisely those factors related to social structure that Merton (1976) emphasizes in his analysis of sociological ambivalence among medical patients. It is not yet possible, however, to determine the relative importance of social structural factors as compared with psychological factors, such as predispositions of the patients making for strong resistance to authoritative demands and idiosyncratic deficiencies of health-care professionals making for insensitive or humiliating treatment of patients.

OBSERVATIONS ON THE SOCIAL POWER OF HEALTH-CARE PROFESSIONALS

Observations on surgical wards and in hospital clinics (Janis, 1958 and 1975; Janis and Leventhal, 1965) have made me keenly aware that the quality of the affiliative bond between patients and health-care professionals is a crucial factor in sustained adherence. Time and again, I have noted that patients will adhere to the rigorous demands of a medical regimen only so long as they have a warm personal attachment to a particular physician, nurse, physical therapist, or other member of the staff. In observing patients suffering from neurological damage to the spinal cord and various other back injuries, for example, I have noticed some of them rapidly becoming hopeless, demoralized invalids. Others, suffering from just as severe disorders, actively struggle against such demoralization. They work day in and day out to carry out all the prescribed procedures and exercises, whether the medical staff is around or not. Sometimes their physicians express amazement over the large number of seemingly lost functions they have been able to restore. What seems to loom large in these successful cases is their warm affectionate relationship with a member of the medical staff, a relationship that bolsters their self-confidence during the most stressful periods of convalescence.

I have also observed similar, though less intensive relationships with the health counselor among the most successful clients who come to anti-smoking clinics and weight-reduction clinics. The clients who perceive the health-care professional not just as a likeable person but also as benevolent, admirable and accepting are most likely to adhere to the recommended "cold turkey" program of smoking cessation or to the low-calorie diet (Janis, in press).

The concept of "referent power" is useful for explicating this facilitating type of affiliative bond between the client or patient and the health-care professional. In the form derived from social psychological analyses of the social power of change agents (French and Raven, 1959; Tedeschi and Lindskold, 1976), the concept refers to the power acquired by professionals when their "acceptance" of clients functions as a major incentive for clients' adherence to prescribed courses of action.

"Referent power" was initially used in a more restrictive sense to describe individuals who fulfill a "comparison reference function," i.e., those used by others as a "frame of reference" for evaluating themselves. With this function in mind, French and Raven (1959) contrasted referent power with other bases of social influence, which are effective in inducing acquiesence but in the absence of surveillance are less likely to create sustained change—coercive power, reward power, and expert power. More recently, social psychologists have extended the concept of "referent power" to apply to influential persons who are regarded as significant others and who have a "normative reference function" (Marwell and Schmitt, 1967; Tedeschi and Lindskold, 1976). Tedeschi and Lindskold observe that normative reference power is a relatively independent dimension of influential persons characterized by high perceived attractiveness and high perceived sociability. Their signs of approval or disapproval function as powerful incentives to induce others to adopt and to internalize their values, attitudes, and standards. This conception of referent power can be derived from the early theoretical analysis of reference groups by Merton and Rossi (1950) and Merton's subsequent elaborations of the social power of reference groups and reference individuals who set and maintain standards for others (Merton, 1957).

When health-care professionals acquire some degree of referent power, they are taken as a "normative frame of reference" with regard to a fairly broad range of health-related attitudes and behavior. If referent power increases to a high level, the professionals' normative influence may extend well beyond this sphere to encompass other aspects of life style.

Physicians and other practitioners, of course, display marked individual differences in the degree to which they rely on the various sources of social power. Years ago, many family physicians apparently developed their referent power to such a high degree that their patients would strive to get well partly because they did not want to disappoint their lovable "old doc." This component is often missing in present-day treatment by specialists (see Merton and Gieryn, 1978). Many physicians and health-care practitioners appear to have little power as reference persons. They rely, in effect, on coercive, reward, legitimate, and expert power, but neglect the potential increase in their ability to influence patients that could come from acquiring referent power as well. These are the practitioners who appear to be most haunted by the spector of nonadherence—of having all their diligence and all their diagnostic and treatment skills come to nothing simply because their patients do not adhere to their recommendations. In contrast, professionals with referent power as well as the other kinds of social power meet with less psychological resistance. Their recommendations are more likely to be internalized and conscientiously adhered to long after the consulting sessions have come to an end. Some of my observations, which will be described shortly, suggest that even when health-care professionals are fully using the other four sources of power, their effectiveness as change agents would increase if they were to adopt one or another means for acquiring social power as reference persons (Janis, in press).

What are the means available to a professional for building up social power as a reference person? The psychological literature on the effects of *positive social reinforcement,* which appears to be a necessary condition for the development of strong affiliative ties, suggests part of the answer (see Bersheid and Walster, 1969). Three widely used types of social reinforcement are part of the folk wisdom of our time for anyone who wants to win friends and influence people: expressing agreement, giving contingent praise, and displaying benevolent interest. A fourth type, much less popular despite the best efforts of Carl Rogers and his followers to make it so, involves giving noncontingent acceptance to bolster the other person's self-esteem. Dittes (1959) has provided experimental evidence indicating that a person's attraction to others in a group rapidly increases when he or she is given accepting comments that raise self-esteem. Similar results were found in a study of pairs of students working together as partners (Jones, Knurek, and Regan, 1973).

Considerable skill, as well as empathy and interpersonal sensitivity, is needed to avoid the pitfalls of using noncontingent acceptance as a means for building up one's referent power. For example, a professional's acceptance statements can have a boomerang effect on his patients if he lays it on so thick that he is presumed to be either habitually insincere or attempting to be ingratiating with hidden manipulative intent (see Jones, 1964). Another type of boomerang effect is illustrated by a little episode (hitherto unpublished) in the life of Nicholas Murray Butler, reported to me by a physician who was an intern at the time Butler was President of Columbia University. Merton (1957, page 381), incidentally, in a discussion of individual differences in the number and complexity of statuses, singled out Nicholas Murray Butler as an extreme example of someone with "enumerable though seemingly endless statuses occupied at the same time." The following episode occurred at a time when Butler was temporarily occupying the unwelcomed additional status of a patient in Columbia University's Presbyterian Hospital. Late one night Butler got up from bed and walked agitatedly down the corridor in his hospital garb, carrying a urine container attached to an indwelling catheter. Noticing his discomfort, the nurse notified the resident physician on night duty, a man who was very keen on expressing a warm, accepting interest in patients. The resident approached Butler with a broad smile and said in his best bedside manner, "What's the trouble, pop, can't you pee?" Coolly looking the physician in the eye, "pop" asked, "Doctor, what is your name?" After being told, he responded, "And my name, doctor, is Nicholas Murray Butler."

CRISES IN THE RELATIONSHIP BETWEEN PROFESSIONALS AND THEIR CLIENTS

In an effort to learn something more about the way professionals build up their referent power—or fail to do so—in their interactions with clients, I extended the scope of my observations to a variety of professional settings where people come for help. I functioned as a professional counselor in various kinds of clinics where clients seek help with marital problems, choosing or changing their careers, giving up smoking, going on a diet, or undergoing disagreeable medical treatments recommended by a physician (Janis, 1975). I met with each client once or twice a week for several weeks, usually from 3 to 12 sessions. In the case of the dieting clinics, the objective of counseling was to help clients carry out a difficult course of action in the face of temptations to backslide; in other clinics, the objective was to help clients arrive at a decision concerning marriage or career by encouraging them to go through the necessary steps of exploring alternatives, seeking pertinent information, and making unbiased appraisals.

After comparing successful and unsuccessful cases in an impressionistic way, I tried to evaluate the plausibility of my inferences from these clinical observations in the light of the clinical and social psychological literature on helping relationships. The main hypotheses that emerge appear to be consistent with Merton's (1957) account of the norm-setting functions of reference groups and reference individuals. They are also consistent with findings from systematic studies indicating that social support from a significant person or group can have positive effects under two main conditions: (1) the relationship is characterized by a high degree of *cohesiveness,* which is determined by the participants' anticipations of socio-emotional gains (such as friendship and esteem) as well as utilitarian gains (such as improved health or better career opportunities) resulting from the relationship with the significant person or group; and (2) the relationship involves exposure to *norm-setting* communications, which convey the behavioral standards that the significant person or group expects one to live up to (Cartwright and Zander, 1968; Hare, 1976; Shaw, 1971).

The hypotheses pertain to three critical phases that typically arise in helping relationships. These hypotheses specify several new variables in addition to the more familiar ones pertaining to social power and positive social reinforcements that I have just discussed—variables often overlooked by many professional practitioners but which can nevertheless affect the extent to which a client or patient will be favorably influenced. The key variables are listed in TABLE 1, organized according to the three critical phases. Most of the variables have already been subjected to experimental investigation, conducted mainly in smoking-cessation clinics and weight-reduction clinics (Janis, in press). I shall summarize the evidence from those studies after giving a short theoretical account of the critical phases.

My observations suggest that three critical phases, arising from three major sources of ambivalence, regularly appear in helping relationships. When these crises are surmounted, people are most likely to benefit from the attempts of a professional to help them arrive at or adhere to difficult decisions.

In the first phase, the practitioner dissipates the patient's wariness and acquires motivating power as a significant "reference person." Fears of being exploited, of being dominated, and of being rejected by the stranger (who is supposed to be a professional helper) constitute initial bases for ambivalence. If the practitioner encourages patients to disclose personal feelings, troubles, or weaknesses and responds to the self-disclosures with statements of noncontingent acceptance, the patients tend to develop an attitude of trust and to rely upon the practitioner for enhancing their self-esteem. The patient's image of the practitioner becomes that of a warm, understanding protective figure who can be counted on to accept personal weaknesses and defects. Normative referent power is also increased if practitioners provide fresh insights or explicit encouragement, regardless of how "awful" the patients characterize themselves.

In the second phase, practitioners begin to use their motivating power. However, the relationship built up during the first phase is impaired as the practitioner begins to function as a norm-sending communicator by encouraging or urging the patient to carry out a necessary but stressful course of action. This is a second major source of ambivalence. If the practitioner makes no such demands, either explicitly or implicitly, the relationship will continue in a warm, friendly way but will be ineffectual. The crisis that arises when practitioners recommend a new course of action are more likely to be surmounted if they make it clear that their demands are very limited in scope and that occasional failures to live up to those demands following a sincere attempt to do so will not change their basic attitude of acceptance toward the patients. It may also be

TABLE 1

CRITICAL PHASES AND ELEVEN KEY VARIABLES THAT DETERMINE THE DEGREE OF
REFERENT POWER OF PROFESSIONAL PRACTITIONERS AS CHANGE AGENTS*

Phase 1 (In order for practitioners to *build up* motivating power)	1. Encouraging a moderate amount of self-disclosure by the client *versus* encouraging little or no self-disclosure. 2. Giving positive feedback (acceptance and understanding) *versus* giving neutral or negative feedback in response to self-disclosure. 3. Using self-disclosures to give insight and cognitive restructuring *versus* giving little insight or cognitive restructuring.
Phase 2 (In order for practitioners to *use* their motivating power)	4. Making directive statements or endorsing specific recommendations regarding actions the client should carry out *versus* abstaining from any directive statements or endorsements. 5. Eliciting commitment to the recommended course of action *versus* not eliciting commitment. 6. Attributing the norms being endorsed to a respected secondary group *versus* not doing so. 7. Giving selective positive feedback *versus* giving noncontingent acceptance or predominantly neutral or negative feedback. (By selective feedback is meant negative feedback in response to any of the patient's comments about wanting to avoid acting in accordance with the recommendations combined with positive feedback in response to all other comments, whether relevant to the decision or not). 8. Giving communications and training procedures that build up self-control attributions and a sense of personal responsibility *versus* giving no such communications or training.
Phase 3 (In order for practitioners to *retain* their motivating power and to *promote* *internalization* after contact ends)	9. Giving reassurances that the practitioner will continue to maintain an attitude of positive regard *versus* giving no such reassurances. 10. Making arrangements for phone calls, exchange of letters, or other forms of communication that foster hope for future contact, real or symbolic, at the time of terminating face-to-face meetings *versus* making no such arrangements. 11. Giving reminders that continue to foster self-control attributions and a sense of personal responsibility *versus* giving no such reminders.

*Based on Janis, in press.

helpful for practitioners to attribute the norms that are being endorsed to a respected secondary group and to negotiate an agreement with the patient so that they become committed to those norms.

Norm-sending practitioners are most likely to retain motivating power when they use a selective pattern of social reinforcement. This pattern consists of criticizing patients' counternorm asertions in a nonthreatening way while expressing positive regard the rest of the time, including when the patients admit to personal shortcomings that are irrelevant to the task at hand. By expressing noncontingent acceptance most of the time and restricting contingent acceptance to the agreed-upon task, practitioners can lead patients to develop an authentic image of them as a quasi-dependable source of self-esteem enhancement; this, in turn, facilitates the practitioners' effectiveness.

A modified expectation of partly contingent acceptance from practitioners allows clients to look forward to receiving genuine acceptance and approval much of the time. The clients sometimes think acceptance will be forthcoming practically all the time, almost as much as when acceptance is in fact wholly unconditional—provided only that they make a sincere effort to follow the relatively few recommended rules which pertain to a limited sphere of personal behavior. The image of a quasi-dependable enhancer differs significantly from that of the garden variety of undependable enhancer who offers acceptance only conditionally. The latter type of helper leads clients to expect little or no approval except when they clearly earn it by conforming to many rules laid down by the helper to govern many spheres of personal behavior, much like the rules imposed by strict parents who demand conformity with their entire code of moral behavior and etiquette. Clients know that signs of acceptance from demanding authority figures are few and far between. An image of the helper as supplying a nurturant diet of variable but basic acceptance, rather than either the meagre bones of conditional acceptance or the rich but undigestible fare of unconditional acceptance, is assumed to be optimal for functioning effectively as a constructive motivator.

In the third critical phase, the influence of the supportive norm-sending practitioner is threatened by the patient's disappointment and resentment centered on the termination of direct contact. As the sessions with the practitioner come to an end, the patient may interpret the termination of contact as a sign of rejection or indifference and fail to internalize the norms advocated by the practitioner. The client's disappointment may become so extreme as to lead to deliberate violations of the counselor's norms. Once a relationship with a helper has been established, aggrievement about termination of contact is a major source of psychological ambivalence. Patients whose chronic diabetes has been stabilized, for example, may stop following the prescribed medical regimen within a few months after they no longer have appointments to see the physician. Adverse reactions to separation may be reduced if the practitioner gives assurances of continuing positive regard and arranges for gradual rather than abrupt termination of contact. In order to prevent backsliding and other adverse effects when contact is terminated, the patient must internalize the practitioner's norms. Little is known about the determinants of this process, but it seems plausible that internalization might be facilitated by communications and training procedures that build up appropriate self attributions and a sense of personal responsibility.

The art of effective health care may require dealing with each of the three critical phases in ways designed to minimize adverse reactions. Perhaps relatively few practitioners have the interpersonal skills that enable them to treat most of their clients with consumate artistry in all three phases. Nevertheless, practitioners having modest amounts of skill in dealing with people in trouble may be able to improve their effectiveness by taking account of the prescriptive hypotheses suggested by this analysis.

The foregoing account of three critical phases assumes that a helper's chances of being effective increase if the client develops a differentiated attitude of reliance on, respect for, and emotional attachment to the professional practitioner or other helper. This is a much more complex attitude than simple "liking" for a stranger as measured by standard scales of interpersonal attraction in current social psychological research (see Berscheid & Walster, 1969; Byrne, 1971). One would expect that after patients have started to reveal personal weaknesses, noncontingent acceptance by the practitioner will result in improving their self regard. This, in turn, will make for strong motivation to continue the relationship, with high reliance on the helper as a respected model. From this point the helper becomes a significant reference person for the client and has the potentiality of using his or her social power to influence the client's actions. This power may compare with that of cohesive normative reference groups. It can be used to encourage the client to become committed to a difficult course of action and to overcome temptations to backslide. The important point is that professional counselors acquire far more social power as normative reference persons and will be regarded with far more affection and deference when clients see them as dependable sources for enhancing self esteem, rather than as conditional acceptors, who are at best undependable enhancers.

A vivid illustration of a client's response to a professional counselor who has become a powerfully motivating norm-setter can be cited from the innovative work of Neal Miller and Barry Dworkin on biofeedback training. These investigators were pioneers in developing an instrumental conditioning technique with verbal rewards to help hypertension patients gain control over their blood pressure. A young woman who wrote down her impressions of an arduous 10-week training period, during which she temporarily succeeded in lowering her diastolic pressure from a dangerously high average of 97 to a satisfactory average of about 80, had this to say about her trainer:

> I always depend very heavily on Barry Dworkin's encouragement and on his personality. I think he could be an Olympics coach. He not only seems aware of my general condition but he is never satisfied with less than my best, and I cannot fool him. I feel we are friends and allies—its really as though *we* were lowering my pressure. (Jonas, 1972.)

When the patient regards the health-care professional as an Olympics coach, she conveys the idea that in some sense she thinks of the coach as treating her like an Olympics star. Not everyone who engages in professional work can expect to function like a successful Olympics coach with all clients. But perhaps a better understanding of the crucial ingredients of an effective helping relationship will lead to improved means for building the type of relationship that is most effective. The key theoretical concepts introduced in the analysis of the three critical phases, including those pertaining to the image of the helper as a quasi-dependable source of self-esteem enhancement, provide a general framework that might account for what happens in a variety of other dyadic relationships—between a student and a teacher, a novice and a guru, a pair of work colleagues, friends, lovers, or marital partners—and also in relationships between group members and their leaders.

RESEARCH ON THE EFFECTS OF BUILDING REFERENT POWER

Only a few of the independent variables listed in Table 1 have been sufficiently investigated to provide replicated findings. Much of the pertinent research carried out in our research program at Yale University deals with the variables in phase 1, the acquisition of referent power (Janis, in press). We have also completed a few studies and have others under way that bear on the variables affecting the other two phases.

One hypothesis that follows from the theoretical analysis of the first phase is that when counselors respond to the self-disclosures with genuine and explicit acceptance, the clients will regard them more positively as a source of self-esteem enhancement. The counselor's recommendations will then be more effective than when they respond in a negative, neutral, or noncommittal way. In a study designed to test the effects of positive vs. neutral feedback, Dowds, Janis and Conolley (in press) carried out a controlled field experiment in the Yale weight-reduction clinic. Overweight women were given a standard interview which was designed to elicit a moderate amount of self-disclosure of personal failures, negative feelings about the self, and other personal information that is normally withheld in everyday conversation. Subjects were assigned on a random basis to two experimental conditions. One condition consistently provided *positive* feedback, that is, the interviewer made explicit comments of approval for cooperating and "acceptance" comments reflecting the key content of the subject's answers, as is done in the Rogerian client-centered style of counseling. The contrasting condition provided only *neutral* feedback; that is, the interviewer made neutral comments or gave no response at all following the subject's answers. After the self-disclosing interview, in both experimental conditions the interviewer became a norm-sending communicator by presenting explicit recommendations concerning adherence to a 1200 calorie diet.

The results of this initial study indicate that positive reinforcement from a counselor, as compared with neutral feedback, when given during an interview that encourages self-disclosure of personal weaknesses, has favorable effects. It fosters more positive ratings of the counselor and also more adherence to the recommended low-calorie diet, as measured by weight loss two months later. Other studies have replicated these basic findings (Chang, 1977; Conolley, Janis and Dowds, in press; Greene, in press.)

Greene's study introduced variations in physical proximity that had not been investigated in the earlier studies. He found that when the seating arrangement for the interview placed the client at a normal distance of about two feet from the counselor, positive verbal feedback had the expected favorable effect, as shown by significantly greater weight loss five weeks after the interview. But when the seating arrangement placed clients at a relatively far distance from the counselor (five feet away), which they apparently interpreted as a nonverbal sign of withdrawal or detachment, the favorable effect of positive verbal feetback was lost. These results point to the same conclusion suggested by subsidiary findings from the earlier experiments in the Yale clinic: In order for positive feedback to be effective, the counselor must use it *consistently* throughout the interview, abstaining from saying or doing anything that could be construed by the clients as withdrawing from them or criticizing them.

Additional confirmatory results, which also indicate limiting conditions, were obtained in a pair of field experiments by Mulligan (in press), conducted during a Red Cross campaign to elicit blood donations from college students. Mulligan found that consistently positive feedback from the interviewer (as compared with neutral feedback) increased adherence to the interviewer's recommendation to donate blood to the Red Cross. This was the outcome when the interview elicited self-disclosures that were not directly relevant to the current decisional conflict. But the opposite outcome was obtained when the interview dealt with the issue of whether or not to donate blood, which gave the clients the opportunity to express their reluctance to do so. These findings indicate that although positive feedback may generally be more effective than neutral feedback in response to a person's self-disclosures, it can be less effective as a result of reinforcing the "wrong" decision from the recommender's standpoint. Even though professionals deliberately attempt to avoid reinforcing "bad intentions," they may inadvertently do so by expressing understanding and empathy during interviews in which clients talk openly about not wanting to carry out a recommended healthful or socially desirable course of action.

In a number of further studies, my coworkers and I have investigated the effects of different levels of self-disclosure by varying the content of the counselor's questions, while holding constant the personality of the counselor, the length of the interview, the recommendations being made, and everything else we could think of that might affect the outcome (Janis, in press). A series of experiments in the Yale weight-reduction clinic and in the Red Cross blood donation campaign support the hypothesis that professional counselors or health-care practitioners will be more effective in inducing adherence to their recommendations if they first elicit a moderate degree—rather than a very low degree—of self-disclosure in the initial session with each of their clients. This is so if they generally give positive feedback in the form of acceptance responses and never display signs of rejection or unfriendliness. (Colten and Janis, in press; Mulligan, in press; Quinlan, Janis, and Bales, in press).

Evidence from an additional set of field experiments, however, does not support the notion that the more disclosure the better. Weight-reduction studies by Quinlan and Janis (in press) and by Riskind and Janis (in press) show that eliciting a relatively high level of self-disclosure results in less behavioral adherence to the counselor's recommendations than eliciting a moderate level of disclosure. The high-disclosure interviews used in these studies—which elicit a great deal of confidential material about the client's weaknesses and shortcomings that seldom, if ever, is disclosed even to intimate friends—are similar to the probing intake interviews used by some "depth" psychologists and psychiatrists who treat people seeking help to control their smoking, eating, drinking, or sexual habits. The findings suggest that such intake interviews are likely to be far less effective in helping clients change their behavior than those that elicit only a moderate amount of personal information.

Since one set of experiments shows that moderate disclosure is more effective than low disclosure and another set shows that high disclosure is less effective than moderate disclosure, the obvious inference is that the relationship between amount of induced disclosure and adherence to the counselor's recommendations is nonlinear. The curve for adherence as a function of self-disclosure would be expected to take the form of an inverted U-shaped function, just as has been found for other types of independent variables that have both facilitating and inhibiting effects. This implication of the findings will have to be tested systematically in parametric studies that vary the amount of self-disclosure within each experiment from very low through intermediate levels to very high.

What accounts for the relatively detrimental effects of high disclosure? Fairly consistent answers are implied by the process data obtained from the previously cited experiments on self-disclosure, and also from the more sensitive indicators obtained in a small sample study by Janis and Quinlan (in press). In the latter study each client was given an intensive process interview immediately after the initial high-disclosure or moderate-disclosure interview. Two main types of detrimental effects were observed. First, numerous signs indicate that participating in a high disclosure interview makes clients somewhat demoralized, despite all the positive comments and acceptance statements of the counselor. After having revealed all sorts of personal weakness, some clients feel dissatisfied with themselves, as well as with the counseling session, and their self-confidence is shaken. When this occurs, the clients feel less certain than ever that they can succeed in carrying out difficult tasks, like adhering to a low-calorie diet. The second type of detrimental effect, suggested by more indirect and subtle indicators, is a relative increase in conflict about entering into a dependent relationship with the counselor. Some clients explicitly express a sense of vulnerability from having revealed so much. Others seem to manifest over-involvement in the emerging dependent relationship by indicating that they really want the counselor to give them more time and more directive advice, not just about the problem at hand (such as being overweight) but also about other problems that were mentioned in the high disclosure interview (such as what to do about marital difficulties).

The clients given a moderate disclosure interview seem less likely to regard the counselor as someone who could become a protective parental figure or a savior who will solve their problems by telling them exactly what to do. At the end of the initial session they appear to accept with more emotional tranquility a business-like relationship with the professional and do not feel deprived because of the limited amount of help offered them. They regard the counselor as friendly, genuinely helpful, and doing a good job.

The various research studies we have just reviewed, including the controlled field experiments and the qualitative analyses of individual case studies, converge on the following conclusion: When a practitioner gives consistently positive feedback, referent power is fostered by eliciting a *moderate* degree of self-disclosure (rather than a very high or low degree) which results in *enhancement of the clients' self esteem* and makes for *increased adherence* to the practitioner's recommendations. There is also evidence in some of the studies indicating that in short-term health counseling, authority figures who remain detached and demanding can also be effective even though they do not enhance self esteem, provided that they are seen by the clients as benign protectors.

An experiment by Langer, Janis and Wolfer (1975) was designed to investigate another factor in the first critical phase (TABLE 1)—cognitive restructuring. The experiment assessed the effectiveness of a reappraisal intervention introduced by a professional counselor to hospitalized patients who had recently decided to undergo a major surgical operation. The cognitive reappraisal procedure consists of developing hope for successful long-term outcomes by encouraging an optimistic but realistic reappraisal of stressful postdecisional events that might otherwise make patients regret their decisions. The procedure proved to be effective in reducing both pre- and postoperative stress, as indicated by both subjective and objective measures. Similar cognitive-reappraisal interventions can probably be developed to prevent postdecisional regret and failure to adhere to prescribed regimens among patients facing other kinds of stressful medical treatments.

Additional evidence from controlled field experiments by Rodin and her colleagues conducted with weight-reduction groups supports the hypothesis that using the clients' self-disclosures to assess and to modify their causal attributions can lead to cognitive reappraisals that facilitate adherence to medical recommendations (Rodin, 1978).

Studies of physician-patient relationships by other investigators provide additional evidence that is consistent with hypotheses about the variables specified in the three phases. For example, in a study of mothers in a rheumatic fever prophylaxis program, adherence to medical recommendations for their children was found to be twice as great among mothers who thought their pediatricians had favorable regard for them than among those who did not (Elling, Whittemore and Green, 1960). Self-esteem enhancement may be especially relevant in pediatric practice since a high percentage of the mothers of ill children have been found to blame themselves and to feel guilty about their children's illnesses—40 percent is the figure reported in a study of 800 mothers by Korsch and co-workers (1968). Another study showed that mothers who openly disclosed some of their fears and emotional tension (other than anger) to the pediatrician subsequently tended to show more adherence to the medical recommendations than mothers who did not (Freeman, 1971).

Several studies of adult patients also supply evidence that is consistent with the theoretical analysis of phase 1 (acquiring motivating power). Adherence to the physician's recommendations was found to be greater when the tone of the physician's comments to patients was positive than when it was negative (Francis, *et al.,* 1969). In the same study adherence was found to be significantly lower when the physician gave essentially neutral feedback after asking the patients to disclose personal information or made no reference to the disclosed information.

Other studies, which bear on phase 3, indicate that weight loss is more likely to be sustained if periodic personal contact with the relevant practitioner is maintained (Brownell, Heckerman and Westlake, 1976). The same has been found also for smoking cessation, even when contact is maintained only by telephone (Shewckuk, 1976).

Although pieces of evidence from a variety of systematic investigations support some of the hypotheses derived from the analysis of the three critical stages, the theoretical assumptions concerning the positive effects of self-esteem enhancement have not yet been fully tested. Nevertheless, the main hypotheses appear to be promising in light of existing studies on professional-client relationships.

LINKAGES OF PSYCHOLOGICAL WITH SOCIOLOGICAL AMBIVALENCE

Upon reading Merton's recent book (1976), I realized that in my studies of psychological sources of ambivalence among the clients of health-care professionals I had taken it for granted that the interactions were at least partly governed by role prescriptions. The structural features of the social relations were taken as "facts of historical circumstance," as Merton puts it (p.4). As I read his analysis of the factors in the structure of social roles that affect the probability of ambivalence arising from the interaction of professionals with their clients, I noticed that some of those factors are similar to the variables listed in my psychological analysis of critical phases. For example, in Merton's account the professional has the authority (a) to induce intimate disclosures of private information, which can have an adverse affect on the client's self esteem (phase 1 in TABLE 1), (b) to prescribe regimens that impose frustrations (see phase 2), and (c) to terminate contact even though the client may want to continue (see phase 3).

In light of Merton's analysis, I am now inclined to view my work as focussing on the psychological ambivalence induced in clients by the behavior of practitioners when they act within the acceptable boundaries of the professional role. The term "boundaries" takes into account the more or less broad range of acceptable ways for practitioners to deal with their clients, including acceptable variations in manners, as well as in the degree to which they use the social power (expert, legitimate, reward and coercive) with which their role is endowed. The resident physician who intercepted Nicholas Murray Butler in the hospital corridor, addressed him as "pop," and asked about his "peeing" was not violating any of the norms governing his professional role. But the role prescriptions gave him options for expressing his interest in an elderly patient on his ward in other acceptable ways that would have had a rather different psychological effect on the patient.

Merton observes in passing that one of the structural attributes of social roles is the "normatively patterned variability in the leeway allowed status-occupants to depart from the strict letter of the norms" (1976, p.16). Another such attribute is "the degree of clarity or vagueness of role-prescriptions" (p.17). These two and a number of other structural attributes contribute to a conception of *allowable leeway* for living up to the role (see Merton, 1968, pp. 371-2). This concept helps to link sociological ambivalence with psychological ambivalence.

From sociological inquiry we can expect to learn about the acceptable range and the central tendency of the norm prescriptions and expectations that comprise a social role. Sociological observations can help us to understand the antecedent societal conditions that give rise to incompatible role prescriptions, and the social consequences of those clashing norms. For example, Merton (1976, pp. 6-8) has described some of the consequences of incompatible normative expectations in contemporary western society that are assigned to the therapist role of the physician, which calls for both a fairly high degree of affective detachment in dealing with patients and a fairly high degree of compassionate concern about them. These incompatible norms give rise to ambivalence in physicians, as manifested by oscillations or compromises in their behavior which, in turn, induce ambivalence in patients.

A different kind of inquiry, which we label as psychological research, investigates other kinds of antecedent conditions (such as personality predispositions and situationally-induced mood states) that give rise to variations in the way a social role is executed. Another type of psychological inquiry treats the behavior of persons who occupy a role as stimuli (or as antecedent conditions) and investigates the effects of variations in their behavior on people with whom they interact, including those variations that are well within the acceptable range for the social role. The psychological variables in TABLE 1, for example, can be regarded as social stimulus variables that describe the behavior of health-care practitioners. They pertain to variations that are well within the acceptable boundaries for the professional role. These variables, according to the psychological hypotheses presented earlier, have the effect of increasing or decreasing the clients' level of self-esteem and the intensity of their ambivalence, which affect the amount of social influence the practitioner can successfully exert.

A "double focus" on sociological and psychological aspects, as recommended by Merton, is essential for a comprehensive analysis of the causes and consequences of ambivalence, which could eventually lead to an integrated "psycho-social theory" (1976, p.6). In the meantime, as he suggests, the two types of inquiry can complement each other.

Paul Lazarsfeld (1975, p.57), commenting on the original sociological ambivalence paper by Merton and his co-author, Elinor Barber, says that it offers a "brilliant program for an empirical study: Why do patients feel uneasy about their physicians?" I would like to add that another research program is also suggested by the same paper—one requiring a combined focus to answer a question about even more profound social consequences: Why do so many patients fail to do what their physicians or other health-care professionals recommend? By using an approach that looks at both the sociological and the psychological sources of ambivalence and their linkages it should be possible to arrive at more complete understanding of the antecedent conditions that give rise to nonadherence.

What kind of hypotheses would link the two types of ambivalence? To sketch out a few examples that might provide some evocative ideas, I shall reformulate the three critical stages specified by the analysis of psychological ambivalence (summarized in TABLE 1) in terms that take account of sociological ambivalence.

From their first contact with clients, physicians and other health-care professionals, by virtue of their social status, have considerable potential for social influence. They wield considerable legitimate and expert power and in some circumstances reward and coercive power as well. There is also, of course, some degree of latent ambivalence right from the outset. When they need the help of a professional, people are wary about the possibility of being exploited or humiliated, sometimes because of prior unhappy experiences with similar professionals, more often because they have heard cautionary tales from friends or relatives. Merton (1976) points out that disgruntled clients give selective or exaggerated accounts to others of the most dramatic failures which "serve to spread ambivalent attitudes and to provide a context for hearers when next they have dealings with professionals" (p.30-31). But clients' suspicions are counteracted to some extent by the credentials of professionals, especially by signs that they are affiliated with a prestigious hospital. In any case, the negative components of clients' attitudes toward the professional are likely to be less strong than the positive components. The negative components of the clients' initially ambivalent attitude, which adversely affect their willingness to adhere to the professional's recommendations, can be expected to increase or decrease depending largely on how the professional behaves in face-to-face sessions.

Professional role prescriptions place numerous constraints on the behavior of health-care practitioners, which prevent them from building up the positive components because the constraints interfere with the acquisition of normative reference power. Consider, for example, the contrasting types of interaction that would be expected to occur when a pair of overweight women agree to use the "buddy system" to help each other stick to a 1200-calorie diet given them by a hospital's dietetics clinic, as compared with when the same two clients have individual sessions with a professional counselor. Partners lack the credentials that give professionals a headstart in expert and legitimate power, but usually without realizing it find it relatively easy to meet the essential conditions for acquiring normative referent power (see phase 1 in TABLE 1). They can plunge into self-disclosing conversations, revealing their current hopes, frustrations, concerns, and personal weaknesses, such as their inability to resist temptations to overeat. Typically they respond to each others' disclosures with signs of acceptance, highlighting their similarities and mutual empathy. But equal partners are likely to do a poor job as norm-senders (see phase 2 of TABLE 1). They usually avoid giving selective social reinforcements that are contingent on adherence and are inclined to share the guilt for deviations from the prescribed diet (Janis, in press). Later on, when the time comes for thinking about termination of the partnership, the partners find it relatively easy to make arrangements that foster hope for future contact and to provide reassurances that they will continue to have positive regard for each other (see phase 3 of TABLE 1). Thus, in terms of the analysis summarized in TABLE 1, each partner will acquire and retain motivating power as a normative reference person, but will fail to use that social power to foster adherence to the diet.

The situation would be entirely different if, instead of forming a partnership, each of the clients had individual sessions with a physician, nurse, or professional health counselor in a weight-reduction clinic that advocates the same 1200-calorie diet. In the professional setting, the client's self-disclosures most likely would be restricted to the business as hand (such as information about eating habits) and the practitioner's response to the disclosures would tend to be neutral and detached, with occasional negative comments indicating disapproval when the client reveals an inability to resist temptations to overeat. (Professional role prescriptions regarding the expression of compassion generally pertain only when patients are in a dire state of physical suffering). Furthermore, tending strictly to business, the professional would be unlikely to take the time to give explanations that might provide insights or cognitive restructuring in response to the client's self-disclosures. To the extent that these features characterize the behavior of professionals, they will fail to build up motivating power as normative reference persons in phase 1. The professionals' social power might actually decrease as a result of their detached impersonal stance and critical comments that lower the self esteem of the clients.

When it comes to norm-sending (phase 2), however, professionals can be counted on to say their piece loud and clear. In accordance with professional role prescriptions, they use whatever social power they have to the fullest extent, laying down the law about the regimen to be followed, giving "doctor's orders." They are also likely to elicit commitment, to give selective social reinforcement (approval for good behavior, disapproval for deviations from the regimen), and to attribute their recommendations to prestigious secondary groups, such as medical research scientists. They are not likely to do anything, however, to build up self-control attributions and a sense of personal responsibility in their clients that would foster adherence to the recommendations in the absence of continued surveillance. Nor are they likely to handle termination of contact (phase 3) in a way that promotes internalization of the recommended course of action.

Insofar as the professional behavior of physicians and other health-care practitioners is characterized by these features, the intensity of psychological ambivalence in their clients can be expected to increase with each contact. Still, when they are feeling miserable and worried about their health, clients have some degree of emotional dependency on the professionals as authority figures to whom they turn for help.

In short, a strictly business-like, no-nonsense style of treatment, which is fostered by professional role prescriptions, can be expected to evoke fears of being humiliated and rejected (phase 1), resentments about being ordered to undergo frustrations (phase 2), and aggrievement about being abandoned after the professional has unilaterally decided that no more appointments are necessary (phase 3). If these reactions occur with a high intensity, the outcome is dysfunctional for attaining the objectives of health-care professionals and of the clients, because the latter will either not come back to complete the series of treatments or will fail to comply with the prescribed regimen.

The objectives of both parties are more likely to be met if the practitioners use in certain ways the *leeway* open to them with regard to how and when the professional role prescriptions are to be lived up to. The intensity of ambivalence induced in clients would be expected to decrease, rather than increase, when a health-care professional's behavior is consistent with the variables (listed in TABLE 1) that are assumed to increase normative referent power. For example, where the medical problem requires little self-disclosure—to set a broken bone, for example—a physician might nevertheless spend an extra few minutes finding out about the patient's worries, since these could be relevant to the regimen that will be prescribed. Even more important, the physician can respond to the patient's disclosures, complaints, and anxious questions in an accepting way, using what the patient says to clear up misunderstandings about the medical problems, which might enable the patient to cope better with suffering and disability. When telling the patient about the recommended course of treatment and the regimen the patient should follow, the physician might avoid reducing the acquired normative reference power by making it clear that the demands are as limited in scope as possible and by encouraging the patient to take responsiblity as a well-informed decision-maker.

Physicians usually do not think of their patients as decision makers. In America, Europe, and other western countries, the patient role is starting to change somewhat, in part as a result of malpractice suits and new requirements of informed consent (see Merton and Gieryn, 1978). Traditionally, however, the patient role has been defined as a passive one. I suspect that a survey would show that most physicians still expect their patients to do whatever they tell them to do without complaint or questioning. But, in fact, patients are active decision-makers. First, they must decide whether to seek medical treatment and from whom. Then they must decide whether to accept the treatment the doctor recommends. After that, they must make a series of decisions, sometimes every day, about the extent to which they are going to follow the rules laid down for them in the recommended medical regimen.

Without violating any professional role prescriptions, physicians can encourage their patients to function as active decision-makers, giving them a sense of freedom of choice among alternative treatments and alternative regimens (see Janis and Rodin, in press). This requires giving clarifying communications, inviting and answering questions, and perhaps also introducing some brief training procedures that are specifically designed to build up self-control and a sense of personal responsibility for carrying out the recommended course of action (see variable 8 in TABLE 1). Physicians who do so are less likely to evoke negative reactions. Fostering a sense of perceived control also counteracts debilitating feelings of helplessness, enabling patients to cope more effectively with whatever stress may arise as a consequence of their medical treatments (see Bowers, 1968; Seligman, 1975). Studies of breast cancer patients, for example, have shown that patients do better, as measured by rate of recovery from surgery, when they have a two-stage surgical procedure, as compared with those who undergo a one-stage procedure (Taylor and Levin, in press). The two-stage biopsy allows time for orderly planning and evaluation prior to surgery or therapy and usually includes active participation of the patient in the decision to resort to surgery.

In a field study, Langer and Rodin (1976) assessed the effects of an intervention designed to encourage elderly nursing-home residents to make a greater number of choices and to feel more in control of day-to-day events. Patients given responsibility for making their own decisions showed significant improvement in alertness and fewer deficits of the sort that are commonly regarded as symptoms of senility. From a physician's blind evaluations of the patients' medical records, it was found that during the 6-month period following the intervention, the "responsible" patients showed a significantly greater improvement in health than the comparable patients in the control group. The most striking follow-up data were obtained in death-rate differences between the treatment groups assessed 18 months after the original intervention: 15 percent in the intervention group in contrast to 30 percent in the control group died (Rodin and Langer, 1977). These findings are in line with Ferrare's (1962) original correlational observation that aged people who were relocated in a new nursing home of their own choice lived longer than those who were sent there without being given any choice.

There are both theoretical and empirical grounds for expecting that practitioners who foster a sense of personal responsibility among their patients, while expressing a basic attitude of positive regard (within the accepted boundaries of their professional role), are likely to induce more sustained adherence. This expectation pertains to preventive measures, such as breast self-examination among women in the age group most vulnerable to breast cancer, as well as to convalescent regimens, such as taking medications and restricting activities among men and women recovering from heart attacks.

Giving patients some degree of control over termination of treatments might also alleviate certain sources of ambivalence. If patients feel ready to terminate the treatment and suspect that it is being unduly prolonged, they may become less resistant if the practitioner takes the trouble to brief them about the alternatives and their consequences before asking them to make a decision about continuing. Or, if patients feel that the practitioner is terminating the treatment too soon, a similar type of briefing might alleviate feelings of aggrievement about being abondoned at a time when help is felt to be still needed. This type of briefing, by embodying the variables listed under phase 3 in TABLE 1, could foster internalization of the practitioner's recommendations after termination of contact. By relinquishing the prerogative to exercise full control over termination of treatment, which is one of the sources of sociological ambivalence, the health-care professional can diminish the likelihood of inducing psychological ambivalence in patients without going beyond the boundaries of acceptable role behavior.

In order to meet the essential conditions for becoming effective change agents, practitioners have to overcome their own inner resistances when dealing with patients most in need of social incentives. For many professionals it goes against the grain to give options and to make lots of favorable comments conveying positive regard to patients who do little more than complain about their troubles and appear to be unwilling or unable to mobilize themselves to do what they obviously ought to do. Health-care practitioners, like almost all other middle- and upper-class people, can be expected to have a sense of social equity that makes them reluctant to violate the norms asserting that social rewards should be given only to those who have earned them or who are likely to reciprocate (see Walster, Walster and Berscheid, 1978). There are other major deterrents as well—the emotional strain of being genuinely empathic toward suffering people, the additional effort needed to acquire and to use the interpersonal skills of an effective change agent, and, most salient of all, the added time required, for which there is apparently no room at all in professional schedules that are already overfilled. Nevertheless, there is reason to expect that more and more practitioners, as they become aware of the demoralizing statistics on nonadherence, will become *quality* oriented, which means that they will have to renounce quantity ambitions, and expend extra effort in accordance with the primary goal of improving the health of each of their patients as far as possible.

In the foregoing sketch of the linkages between sociological and psychological ambivalence, I have tried to indicate that nonadherence and other adverse outcomes need not be regarded as inescapable consequences of the structure of professional roles. Practitioners have sufficient leeway to act in ways that meet the essential conditions for acquiring a high degree of referent power as change agents. As evidence concerning those conditions accumulates, it should be possible to develop the comprehensive type of psycho-social theory to which Merton has encouraged behavioral scientists to aspire. It should then not be too difficult to work out and test prescriptive hypotheses that have socially beneficial applications for helping people in trouble. With increased understanding of both sources of ambivalence, behavioral scientists at long last could have sufficient expert power to supply valid prescriptions for increasing the effectiveness of all those who have legitimate power to prescribe.

REFERENCES

ATTHOWE, J. 1973. Behavior innovation and persistence. *Am. Psychol.* **28:** 34–41.

BERGMAN, A. B. & R. J. WERNER. 1963. Failure of children to receive penicillin by mouth. *N. Engl. J. Med.* **268:** 1334–38.

BERSCHEID, E. & E. H. WALSTER. 1969. *Interpersonal attraction.* Addison-Wesley. Reading, Mass.

BOWERS, K. G. Pain, anxiety, and perceived control. *J. Consult. Clin. Psychol.* **32:** 596–602.

BROWNELL, K. D., C. L. HECKERMAN & R. J. WESLAKE. December 1976. Therapist and group contact as variables in the behavioral treatment of obesity. Paper presented at the annual meeting of the Association for the Advancement of Behavior Therapy, New York. N.Y.

BYRNE, D. 1971. *The attraction paradigm.* Academic Press, New York.

CARTWRIGHT, D. & A. ZANDER. (Eds.). 1968. *Group dynamics: Research and theory.* (3rd. ed.). Harper and Row, New York.

CHAMEY, E., R. BYNUM & D. ELDREDGE. 1967. How well do patients take oral penicillin? A collaborative study in private practice. *Pediatrics.* **40:** 188–195.

CHANG, P. 1977. The effects of quality of self-disclosure on reactions to interviewer feedback. Unpublished doctoral dissertation. University of Southern California.

COLTEN, M. E. & I. L. JANIS. Effects of Self-Disclosure and the Balance-Sheet Procedure. *In: Counseling on Personal Decisions.* I. L. Janis, Ed. Yale University Press, New Haven, Conn. (In press)

CONOLLEY, E., I. L. JANIS & M. DOWDS. Effects of variations in the type of feedback by the counselor. *In: Counseling on personal decisions: Theory and research on short-term helping relationships.* I. L. Janis, Ed. Yale University Press, New Haven, Conn. (in press).

DAVIS, M. S. 1966. Variations in patients' compliance with doctors' orders: Analysis of congruence between survey responses and results of empirical investigations. *J. Med. Educ.* **41:** 1037–1048.

DAVIS, M. S. 1967. Discharge from hospital against medical advice: A study of reciprocity in the doctor-patient relationship. *Soc. Sci. Med.* **1:** 336.

DAVIS M. S. 1971. Variations in patients' compliance with doctors' orders: Medical practice and doctor-patient interaction. *Psychiatry Med.* **2:** 31-54.

DITTES, J. E. 1959. Attractiveness of group as function of self-esteem and acceptance by group. *J. Abnorm. & Soc. Psychol.* **59:** 77-82.

DOWDS, M., I. L. JANIS & E. CONOLLEY. Effects of acceptance by the counselor. *In: Counseling on personal decisions: Theory and research on short-term helping relationships.* I. L. Janis, Ed. Yale University Press, New Haven, Conn. (In press).

ELLING, R., R. WHITTEMORE & M. GREEN. 1960. Patient participation in a pediatric program. *J. Health Hum. Behav.* **1:** 183-191.

FEINSTEIN, A. R., H. F. WOOD, J. A. EPSTEIN, A. TARANTA, R. SIMPSON & E. TURSKY. 1959. A controlled study of three methods of prophylaxis against streptococcal infection in a population of rheumatic children. *N. Engl. J. Med.* **260:** 697.

FERRARE, N. A. 1962. Institutionalization and attitude change in an aged population. Unpublished doctoral dissertation, Western Reserve University, Cleveland, Ohio.

FRANCIS, V., B. M. KORSCH & M. J. MORRIS. 1969. Gaps in doctor-patient communications: Patients' responses to medical advice. *N. Engl. J. Med.* **280:** 535.

FREEMAN, G. 1971. Gaps in doctor-patient communication: Doctor-patient interaction analysis. *Pediatr. Res.* **5:** 298-311.

FRENCH, J. R. & B. RAVEN. 1959. The bases of social power. *In: Studies in Social Power.* D. Cartwright, Ed.:150-167. University of Michigan, Ann Arbor.

GREENE, L. Effects of the counselor's verbal feedback, interpersonal distance and clients' field dependence. *In: Counseling on personal decisions: Theory and research on short-term helping relationships.* I. L. Janis, Ed. Yale University Press, New Haven, Conn. (In press.)

HARE, A. P. 1976. *Handbook of small group research.* Second edition. Free Press, New York.

HUNT, W. A. & J. D. MATARAZZO. 1973. Three years later: recent developments in the experimental modification of smoking behavior. *J. Abnorm. Psychol.* **81:** 107-114.

JANIS, I. L. 1958. *Psychological stress: Psychoanalytic and behavioral studies of surgical patients.* John Wiley & Sons. New York.

JANIS, I. L. 1975. Effectiveness of social support for stressful decisions. *In: Applying social psychology: Implications for research, practice, and training.* M. Deutsch and H. Hornstein Eds. Lawrence Erlbaum Associates, Hilldale, N.J.

JANIS, I. L. 1976. Preventing dehumanization. *In: Humanizing Health Care.* J. Howard & A. Strauss, Eds. Wiley & Sons, New York.

JANIS, I. L. (Ed.) *Counseling on personal decisions: Theory and research on short-term helping relationships.* Yale University Press, New Haven, Conn. (In press.)

JANIS, I. L. & H. LEVENTHAL. 1965. Psychological aspects of physical illness and hospital care. *In: Handbook of Clinical Psychology.* B. Wolman, Ed. McGraw-Hill, New York.

JANIS, I. L. & D. M. QUINLAN. What disclosing means to the client: Comparative case studies. *In: Counseling on personal decisions: Theory and research on short-term helping relationships.* I. L. Janis, Ed. Yale University Press, New Haven, Conn. (In press.)

JANIS, I. L. & J. RODIN. Attribution, control and decision-making: Social psychology in health care. *In: Health psychology.* G. C. Stone, F. Cohen and N. E. Adler, Eds. Jossey-Bass, San Francisco. (In press.)

JONAS, G. 1972. Profile: Visceral learning I. *New Yorker Magazine.*

JONES, E. E. 1964. *Ingratiation: A social psychological analysis.* Appleton-Century-Crofts, New York.

JONES, S. C., D. A. KNUREK & D. T. REGAN. 1973. Variables affecting reactions to social acceptance and rejection. *J. Soc. Psychol.* **90:** 264-284.

KASL, S. V. September, 1975. Issues in patient adherence to health care regimens. *J. Hu. Stress.:* 5-17.

KORSCH, B. M., E. K. GOZZI & V. FRANCIS. 1968. Gaps in doctor-patient communication: Doctor-patient interaction and patient satisfaction. *Pediatrics.* **42:** 855-871.

LANGER, E. J., I. L. JANIS & J. A. WOLFER. 1975. Reduction of psychological stress in surgical patients. *J. Exp. Soc. Psychol.* **11:** 155-165.

LANGER, E. J. & J. RODIN. 1976. The effects of choice and enhanced personal responsiblity for the aged: A field experiment in an institutional setting. *J. Pers. Soc. Psychol.* **34:** 191–198.

LAZARSFELD, P. F. Working with Merton. *In: The Idea of Social Structure: Papers in Honor of Robert K. Merton.* L. A. Coser, Ed. Harcourt, Brace, Jovanovich, New York.

LEY, P. & M. S. SPELMAN. 1965. Communications in an out-patient setting. *Br. Jr. Soc. Clin. Psychol.* **4:** 114–116.

MADDOX, G. L., C. G. ANDERSON & M. D. BOGDONOFF. 1966. Overweight as a problem of medical management in a public out-patient clinic. *Am. J. Med. Sci.* **252:** 394.

MARWELL, G., & D. R. SCHMITT. 1967. Dimensions of compliance-gaining behavior: An empirical analysis. *Sociometry.* **30:** 350–364.

MERTON, R. K. 1957 and 1968. *Social theory and social structure.* Free Press, New York.

MERTON, R. K. 1976. *Sociological ambivalence and other essays.* Free Press, New York.

MERTON, R. K. & T. F. GIERYN. 1978. Institutionalized altruism: The case of the professions. *In: Sociocultural change since 1950.* T. L. Smith & M. S. Das, Eds. Vikas, New Delhi.

MERTON, R. K. & A. K. ROSSI. 1950. Contributions to the theory of reference group behavior. *In: Continuities in Social Research.* R. K. Merton and P. F. Lazarsfeld, Eds. Free Press, New York.

MULLIGAN, W. Effects of induced self disclosure and interviewer feedback on compliance: A field experiment during a Red Cross blood donation campaign. *In: Counseling on personal decisions: Theory and research on short-term helping relationships.* I. L. Janis, Ed. Yale University Press, New Haven, Conn. (in press).

QUINLAN, D. M. & I. L. JANIS. Unfavorable effects of high levels of self disclosure. *In: Counseling on personal decisions: Theory and research on short-term helping relationships.* I. L. Janis, Ed. Yale University Press. New Haven, Conn. (In press.)

QUINLAN, D. M., I. L. JANIS & V. BALES. Effects of self-disclosure and frequency of contact. *In: Counseling on personal decisions.* Yale University Press, New Haven, Conn. (In press)

RISKIND, J. & I. L. JANIS. Effects of negative self disclosure and approval training procedures. *In: Counseling on personal decisions: Theory and research on short-term helping relationships.* I. Janis, Ed. Yale University Press, New Haven, Conn. (In press.)

RODIN, J. September, 1978. Cognitive-behavioral strategies for the control of obesity. Paper presented at Conference on Cognitive-behavior Therapy: Applications and issues. Los Angeles, Calif.

RODIN, J. & E. LANGER. 1977. Long-term effects of a control-relevant intervention with the institutionalized aged. *J. Personal. Soc. Psychol.* **35:** 897–902.

SACKETT, D. L. & R. B. HAYNES. 1977. *Compliance with therapeutic regimens.* Johns Hopkins Press, Baltimore, Md.

SHAW, M. E. 1971. *Group Dynamics.* McGraw-Hill, New York.

SELIGMAN, M. E. P. 1975. *Helplessness.* Freeman, San Francisco, Calif.

SHEWCHUK, L. A. 1976. Special report: Smoking cessation programs of the American Health Foundation. *Prev. Med.* **5:** 454–474.

STAMLER, J., J. A. SCHOENBERGER & H. A. LINDBERG. 1969. Detection of susceptibility to coronary disease. *Bull. N.Y. Acad. Med.* **45:** 1306.

TAYLOR, S. & S. LEVIN. The psychological impact of breast cancer: Theory and practice. *In: Psychological Aspects of Breast Cancer.* A. Enelow, Ed. Oxford University Press, London. (In press.)

TEDESCHI, J. T. & S. LINDSKOLD. 1976. *Social Psychology: Interdependence, interaction, and influence.* John Wiley & Sons, New York.

VINCENT, P. 1971. Factors influencing patient non-compliance: A theoretical approach. *Nurs. Res.* **20:** 509.

WALSTER, E., G. W. WALSTER & E. BERSCHEID. 1978. *Equity: Theory and research.* Allyn and Bacon, Boston, Mass.

ZOLA, I. K. 1973. Pathways to the doctor—from person to patient. *Soc. Sci. Med.* **7:** 677.

INTERPRETATION AND THE USE OF RULES:
THE CASE OF THE NORMS OF SCIENCE

Michael Mulkay

Department of Sociology
University of York
Heslington, York, YO1 5DD, England

No discussion of the normative structure of science can begin without immediate reference to the work of Robert Merton. For it is common knowledge that Merton provided the first systematic and the most influential attempt by a sociologist to identify the main norms operative among scientists and to show how these norms contribute to the advance of scientific knowledge.[1] In recent years, numerous studies have been published offering evidence for the operation of the Mertonian norms,[2] describing their historical development,[3] analyzing their effectiveness as the major source of social control in science,[4] and attempting to show that they constitute an ethical code for practicing scientists.[5] At the same time, several attempts have been made to argue that the customary analysis of norms in science is based on misleading assumptions, that its empirical foundation is weak and that it is in need of radical conceptual revision.[6–8] This present paper is an attempt to contribute to this lively, ongoing debate.

Hidden Interpretation of Rules

The central issue which I will try to address is that of the relationship between norms or rules and social action. It seems to me that one of the crucial errors prevalent among sociologists of science, whether they have supported or criticized the Mertonian position, has been to assume that this relationship is relatively unproblematic. In other words, most of us have assumed that, once we have identified the rules which scientists use, we can apply the rules to particular acts without any further interpretative work by the analyst. Unfortunately, in so doing, we have failed to notice a fundamental point made by Wittgenstein, namely, that no rule can specify completely what is to count as following or not following that rule.[9] Sociologists of science have simply not realized that, in arguing for or against the operation of particular norms, they have engaged in hidden interpretation of these norms in ways which support their own case, yet which can in every instance be forcefully challenged.

I will give just two examples. In order to show my "conformity to the rule of impartiality," I will take the first of these from one of my own early papers.[10] In offering an account of nonconformity to the Mertonian norms in the Velikovsky case, I wrote as follows:

Numerous examples can be cited of violation by members of the scientific community of the Mertonian norms. In February, 1950, severe criticisms of Velikovsky's work were published in *Science News Letter* by experts in the fields of astronomy, geology, archaeology, anthropology and oriental studies. None of these critics had at that time seen *Worlds in Collision*, which was only just going to press. These denunciations were founded upon popularized versions published, for example, in *Harper's, Reader's Digest* and *Collier's*. The author of one of these articles, the astronomer Harlow Shapley, had earlier refused to read

111

0028–7113/80/0039–0111 $1.75/2 © 1980, NYAS

the manuscript of Velikovsky's book because Velikovsky's "sensational claims" violated the laws of mechanics. Clearly the "laws of mechanics" here operate as norms, departure from which cannot be tolerated. As a consequence of Velikovsky's non-conformity to these norms Shapley and others felt justified in abrogating the rules of universalism and organized skepticism. They judged the man instead of his work and in this way failed to live up to the demands of organized skepticism, for they did not subject Velikovsky's claims to rigorous examination before assessing the validity of these claims. (pp. 32–33)

The main fault here is that no consideration is given by the analyst to the possibility that the rules in which he was interested could have been given a meaning different from that which he assumed, in relation to features of the context which he chose to ignore. More generally, there is a failure to recognize that in applying rules to specific acts, further variable processes of reasoning are typically involved which are in no way specified in the rules themselves. By varying these processes of reasoning, which remain implicit in the text above, participants or other analysts could quite easily produce alternative rule-based accounts which displayed conformity to rather than deviation from the norms of universalism and organized skepticism as customarily formulated. For instance, it could be argued that the kind of qualitative, documentary evidence used by Velikovsky had been shown time and time again to be totally unreliable as a basis for impersonal scientific analysis and that to treat this kind of pseudo-science seriously was to put the whole scientific enterprise in jeopardy. In this way scientists could argue that their response to Velikovsky was an expression of organized skepticism and an attempt to safeguard universalistic criteria of scientific adequacy. This kind of argument could be extended much further if we so wished and made much more detailed. It is difficult to see how the analyst could conclusively refute (or establish) such an argument without much more evidence than is usually obtained by sociologists; and it may well be that such arguments are not refutable in principle, owing to the open-ended character of rules.

The reader should not infer from this example simply that those who have sought to challenge the Mertonian norms have sometimes been careless. The example chosen above was selected in order to illustrate a general point about the analysis of norms in the sociology of science. To rephrase this point, it is that we should not assume that any norm can have a single literal meaning independent of the contexts in which it is applied. Failure to recognize this easily leads to analysis which is more or less vacuous. It becomes too easy for the analyst to classify virtually any professional act in relation to a supposed norm and to claim, unjustifiably, that the act in question is thereby explained, or that it is thereby shown to be deviant. This kind of analytical weakness pervades the sociological work on the norms of science. It is found even in the most careful and systematic analyses. We can observe it, for example, in a recent paper by Zuckerman, containing perhaps the best exposition of the Mertonian position, where she examines the writings of those who have been critical of that position.

[T]he social institution of science provides a composite of incentives and punishments directed towards having scientists adhere to both the moral and the cognitive norms. Yet some sociologists of science have argued that scientists are not significantly oriented towards the moral norms which, in any case, are not institutionalized through the system of social control. One cannot help but notice that the very statement of that view is interestingly self-disconfirming. It represents conformity to the norms prescribing organized skepticism (critical responses), communism (public communication), universalism (with the criticism being centered on impersonal theoretical issues rather than being ad hominem) and, it appears evident, disinterestedness. (p. 128)

This quotation can be used to illustrate a point quite different from that intended by its author. For in this passage the norms and their power to regulate conduct are simply *taken for granted*. What the analyst has done is to classify certain routine academic acts in terms of the Mertonian norms and then to assert, without further evidence, that these acts obviously represent conformity to those norms. But clearly there is nothing presented in this passage (or elsewhere in her discussion) that prevents us from replacing Zuckerman's hidden analysis with a different interpretation. Why do we have to accept, for instance, that writing a critical article is necessarily due to conformity to a norm rather than to self-interest coupled with perception of some technical inadequacy in the relevant literature? Why is it so obvious that analysts and participants (in this case the participants are the analysts) must regard such articles as disinterested? It would be just as plausible to maintain that one or both parties to the debate were acting so as to preserve their vested interest in the defense of a particular intellectual position; and one can well imagine that, in social contexts other than that of writing a formal paper, such an alternative interpretation would be quite acceptable and seem quite convincing. (One can, indeed, easily read Zuckerman's comment on disinterestedness as ironic rather than literal). Similarly, the absence of what are taken to be *ad hominem* arguments in the literature may just as well be due to the conventions of academic publication than to commitment by participants to a context-free norm of universalism.[8] I suggest, therefore, that Zuckerman, like Mulkay in the previous example, drastically oversimplifies the task of applying rules to specific actions and fails to appreciate the extent to which such application involves inferential work on the part of the analysts and participants.

The failure of sociologists of science to recognize the interpretative character of rules has prevented them from undertaking any detailed analysis of the moral language of science in its full complexity, subtlety and diversity. The basic Mertonian norms, although helpful as a point of departure, have been used in such a loose and all-inclusive way by us all, that any conceivable professional act on the part of research scientists has been classifiable, with a little ingenuity, within their frame of reference. Accordingly there has developed a strong tendency to see the advance of scientific knowledge as depending on a rather mechanical reproduction of "literal versions" of these norms.

> Universalism, organized skepticism, communality, and disinterestedness together provide a context in which maximum progress can be made in scientific advancement. These norms function positively for the development of science through directing behavior in the social interaction required because, science, as much as anything, is a social institution. All social institutions must have rules for members to live by and patterns of behavior that others can depend on and can follow to avoid having to decide at every turn what is the right thing to do.[2] (p. 8)
>
> The goal of science then is "public knowledge." And it is the "scientific community" which pursues this goal. Now the interaction of numbers of individuals in the pursuit of a common goal requires a set of rules to regulate their activities and interactions. . . Given the institutionalized goal of science then, the norms follow as necessary rules for the effective pursuit of the goals.[11] (p. 15)

As these quotations illustrate, sociologists of science have tended to assume that the "institutionalized norms" of science provide pre-established solutions to problems of choice and social evaluation; and that these norms thereby reduce to a minimum the need for interpretative work by participants. However, the evidence produced to support this kind of analysis is rather indirect and far from conclusive on this point; for instance, one of the main empirical findings offered in the first text quoted above is that the number of citations to the papers of a given collection of scientists is more highly correlated with the number of papers than with any other easily quantifiable variable. One can, of course, interpret this as supporting the idea that scientists allocate rewards in accord with a norm of universalism and that effective universalism presupposes the operation of the other Mertonian norms. But it seems to me that this approach tells us very little about the actual social world of science. It takes for granted that the supposed norms of science operate in a literal fashion and that their operation will find direct expression in relatively crude statistical relationships.

As I have suggested above, it is only possible to maintain such an approach by engaging in covert interpretation of the rules. Thus in the kind of citation study mentioned above, the meaning of the norm of universalism and, indeed, the whole "scientific ethos," is defined in effect by that range of particularistic variables that the analyst has been able to measure and "control." This approach, then, prevents us from observing the interpretative work done by scientists. It obscures any variability in participants' interpretations by replacing them with speculative interpretative work performed by the analyst. The strategy I wish to recommend instead is that of trying to identify as directly as possible the full range of rules that scientists actually use in practice and to obtain as much detailed information as possible on the kind of interpretative work carried out by scientists themselves. When this strategy has been adopted in the study of rules in other areas of sociological inquiry, it has tended to lead to marked changes in the kind of analysis produced. What I intend to do now, therefore, is to broaden the scope of the discussion by looking briefly at the findings of three studies dealing with similar issues in other substantive areas, in the hope that a less parochial outlook than is usual in the sociology of science will prove to be analytically beneficial.

Some Studies of the Use of Rules

In recent years, a few empirical studies have been carried out of the ways in which particular rules are used within specific, small-scale social groups. Wootton,[12] for example, studied interaction within therapeutic groups in a hospital; and his analysis illustrates some of the complexities encountered by patients and staff in using the rule "Share your experiences with other patients." One of Wootton's basic conclusions is that, because "organizational rules and rights do not contain instructions which fit them to each instance," such rules "have to be made relevant in some way to instances, and this requires work of various kinds on the part of speakers and hearers. A specification of the rules themselves is then best considered alongside the other interactional work being done to authorize the relevance of the rules to this or that instance. . . ." (p. 341)

In the main body of his analysis, Wootton shows how participants, in order to use and apply the rule of sharing, engage in various supplementary interpretative procedures. For example, he shows how they use various social categories to establish similarities or differences among those to whom the rule might be applied and how such categorization is employed to establish the relevance or meaning of the rule of sharing in specific instances. He also draws attention to the ways in which participants refer in interpreting and applying the rule to the organization of conversational topics, the body of conventional psychiatric knowledge and formal and informal organizational practices. He stresses, however, that the analyst cannot provide an exhaustive list of items which participants regard as relevant to the application of the rule, "because such a list would be indefinite." (p. 347) In short, Wootton concludes that the interpretative work carried out by participants in connection with the rule of sharing is not just complex, but also open-ended. The rule appears to have no stable, literal meaning which covers a defineable class of instances to which it can be applied. It seems most appropriate, therefore, to conceive of the rule as a symbolic resource, the meaning of which is established on each occasion of use through combination with other resources in the course of interaction. If such a view of rules were to be generally applicable, it would clearly require us to modify the kind of normative analysis customary in the sociology of science.

A similar line of analysis is developed by Zimmerman,[13] in a study which goes further than that of Wootton in beginning to specify how rule-use is affected by the context of social interaction. Zimmerman argues that we cannot adequately treat "rules as idealizations, possessing stable operational meanings invariant to the exigencies of actual situations of use, and distinct from the practical interests, perspective, and interpretative practices of the rule user." (p. 223) He illustrates this thesis in relation to the rule "First come, first served," as used by the receptionists of a public assistance organization to assign applicants to caseworkers. It was the practice among receptionists to formalize this rule by means of a caseworkers' interview matrix, in such a way that applicants could be assigned to the appropriate interviewer by being allocated, as they entered, to the next empty square on the grid. This is a particularly interesting case, therefore, because a clear literal application of the rule was provided by the formal matrix. One can reasonably assume that if complex interpretative work was required to implement this rule in this small-scale context, then such work must be absolutely essential in the implementation of the more general formulations which are thought to make up the regulative code operative within the international community of research scientists.

Zimmerman shows clearly that receptionists did not apply the "next available cell" rule literally, i.e. they did not apply it irrespective of variations in the actual situation of use. He gives three examples of how the rule was "suspended, modified or interpreted" in practice. In the first place, there were situations where, because one official was taking much longer to interview applicants than were his colleagues, certain applicants were kept waiting much longer than others. This produced a backlog of applicants who were not being moved smoothly through the system and who tended to become restive. In such situations receptionists sometimes intervened and reallocated one or more of those applicants who had been unduly delayed. Receptionists were aware that such an action could be seen as a departure from the basic rule of allocation. But they were able to argue quite plausibly that reallocation in such circumstances was more consistent with the "intent" of the rule, i.e. with the practical objective of moving applicants through the process with reasonable speed and a minimum of difficulty, then a pedantic, literal interpretation would have been. In other words, the meaning of the rule in such instances depended on the way in which receptionists viewed the rule in relation to their immediate practical objectives and on the way in which receptionists judged that those objectives were or were not being satisfactorily realized.

Secondly, receptionists sometimes allowed applicants to choose their own case-worker, instead of being allocated to the next available space on the matrix. Once again, Zimmerman notes, receptionists tended to link this interpretation to their practical objectives; they presented it as a permissiveness which was necessary in order to avoid difficulties and in order to get on with the work. Furthermore, they described such instances as "one time only" concessions, which should not be regarded as precedents and which did not, therefore, undermine the rule-guided character of their normal procedures.

Thirdly, some caseworkers were able to establish a right to deal with potentially troublesome applicants or those needing out of the ordinary treatment. Thus both applicants and caseworkers "interfered" with the literal application of the rule of allocation in ways which were not explicitly formulated. It is quite clear, then, that the receptionists' formal matrix alone did not govern the allocation of applicants to caseworkers and that the literal "First come, first served" rule was not the sole basis of receptionists' action in assigning new entrants.

In the course of his discussion of this material, Zimmerman[13] concludes that we must find an alternative to "the compliance model of rule use typically employed in sociological studies of rule-governed behavior." (p. 225) In trying to understand how Zimmerman reaches this conclusion, it is useful to consider whether one could modify the unduly simple compliance model of receptionists' action, based on literal conformity to one rule, which provides his point of departure. In particular, we could try to devise a set of supplementary, informal rules dealing with exceptional cases, which receptionists use when the basic rule is inapplicable owing to the presence of some unusual feature. The major difficulty with this response, however, has already been mentioned above by Wootton, namely, that the list of such features seems likely to be indefinite and not open to formalization. What seems to happen at the reception counter is that receptionists respond to individual applicants with subtle and tacit judgments of how troublesome they are likely to be, how much they are likely to interfere with the smooth operation of the system. Accordingly, in principle, any feature which is perceived as an actual or potential source of trouble or disruption can provide the basis for reinterpretation of the rules; and it is hardly possible to identify all such features in advance. Furthermore, such features will not be stable characteristics of applicants as such, but will grow out of and change with the course of interaction between applicants, receptionists and caseworkers. It is difficult to see how these judgmental processes could be regulated by formal rules; for they would have to resolve such intangible issues as whether an applicant seemed sufficiently troublesome to warrant the application of one or more supplementary rules, or whether an applicant had or had not been waiting "too long." It is important to realize that if we were to introduce a series of supplementary, informal rules into the analysis and to recognize the necessity of informal judgments, this would have important implications for our conception of routine cases, i.e. those which appeared to fall unproblematically under the basic rule of allocation. For, if receptionists employ a range of formal and informal rules rather than one literal rule, then interpretative work must be involved in identifying routine cases as well as exceptional cases. The allocation of an applicant to the next empty box necessarily implies that the receptionist has judged it inappropriate to bring into play any of the "supplementary rules;" and such judgments cannot be deduced directly from any set of explicit, pre-established rules.

In view of these kinds of consideration, Zimmerman decides that it is inappropriate to conceive of routine social interaction as proceeding simply in conformity with socially internalized and sanctioned rules. He abandons the traditional notion that institutionalized rules operate as standardized solutions to problems of choice or judgment. Choice is in fact still very much involved. For any rule or set of rules is compatible with a great, and probably indefinite, variety of actions which cannot be derived from the rule(s) in advance by means of any formal procedures, because they depend on the interpretative work carried out by participants in specific social contexts. Accordingly Zimmerman suggests that we, as sociologists, should not focus on the issue of participants' compliance with our understanding of the rules, but upon the ways in which participants themselves employ and give meaning to their rules.

[T]he notion of action-in-accord-with-a-rule is a matter not of compliance or noncompliance per se but of the various ways in which persons satisfy themselves and others concerning what is or is not "reasonable" compliance in particular situations. Reference to rules might then be seen as a commonsense method of accounting for or making available for talk the orderly features of everyday activities, thereby making out these activities as orderly in some fashion. Receptionists, in accomplishing a for all practical purposes ordering of their task activities by undertaking the "reasonable" reconciliation of particular actions with "governing rules" may thus sustain their sense of "doing good work" and warrant their further actions on such grounds.[13] (p. 233)

In short, rules are formulations which are used by participants to create or present particular versions of the ordered social reality in which they are engaged. By showing that particular actions follow naturally from the letter or intent of a rule, participants attach to those actions a sense of "this is how it ought to be," a sense of moral necessity or of conventional propriety. In so doing, they are able to bolster their own feeling of having acted correctly, to justify actions which are challenged, and to provide for the regular reproduction of what can be taken as a "normal and proper state of affairs."

The final study to be discussed here is that by Wieder of the convict code. The convict code is a set of rules or a "verbalized moral order" which appears to be generally endorsed among convicts, parolees and residents of rehabilitative organizations. It includes such maxims as, "Do not inform on other residents," "Do not admit that you have done anything wrong," "Do not trust staff," "Share what you have with other residents," "Do not interfere with other residents' interests," and so on.

> Essentially the same code has been found in a variety of settings and has been utilized by sociologists to account for a variety of deviant behavior: violation of institutional rules, refusal to give information to officers, hostile gestures and talk towards officers, threats against officers, gambling, stealing from the institution, sharing stolen goods, engaging in homosexuality, avoidance of contact with staff, and avoidance of participation in group therapy programs. These behaviors have been traditionally analyzed as produced by compliance to the convict code.[14] (p. 125)

Thus the content of the code, its internalization by participants, and its support by positive and negative sanctions, have been seen by sociologists as providing the basis for an adequate explanation of a wide range of recurrent actions within the criminal subculture.

Wieder argues, however, that the customary normative explanation of convicts' actions is unsatisfactory for two main reasons. In the first place, he shows that on the basis of his own and other sociologists' observations of convicts, parolees and residents, it is possible to formulate, not just one set of rules, but various "plausible and competitive sets." (14:197) Similarly, although the rules of the convict code can be made to appear consistent with observed actions after the event, these rules do not tell the investigator specifically what to expect on any particular occasion.

> If some of the behaviors actually proposedly encompassed by [a] rule are examined, the nonpredictability of the behaviors is . . . dramatic. From the rule, "Show your loyalty to the residents", how should an analyst propose that residents would sit at group? Would they be "tense and hostile" in their posture, or would they be so relaxed as to appear disinterested? How should he expect them to respond to requests put forth by the staff? Would they be very resistant to direct orders and less resistant to permissively given suggestions that they do something, or would they undertake no action unless they were "forced" to do so? In both of these instances, either chosen alternative would be equally plausible interpretations in terms of the same rule, even though the alternatives propose opposite actions.[15] (pp. 167–8)

Thus the relationship between actions and rules is indeterminate and incapable of furnishing the kind of deductive explanation which sociologists have traditionally sought. This does not mean, of course, that it is never possible to treat observed actions as necessary consequences of literal conformity to a code. Indeed, Weider stresses that this is precisely what participants themselves typically do. He shows in considerable detail how participants classified actions in terms of the code, interpreted motives in accordance with the code, and generally created the appearance that action was normatively structured by the code in a determinate way. "Somehow through the vehicle of ordinary conversation, residents made it happen that their behavior would be seen as regular, independent of their particular doing, and done as a matter of normative requirement."[14] (p. 146) The code was used to identify "the meaning of a resident's act by placing it in the context of a pattern. An equivocal act then becomes 'clear' in the way that it obtains its sense as typical, repetitive, and more or less uniform, i.e. its sense as an instance of the kind of action with which staff was already familiar. Staff's environment was also structured by the flexibility of 'telling the code,' which could render nearly any equivocal act sensible in such a way that it was experienced as something familiar, even though the act might not be 'expected' or 'predicted' in any precise meaning of those terms."[15] (pp. 149–150)

The fact that participants continually characterize actions in terms of the code and as necessary consequences of the code creates the second major difficulty for the customary form of sociological analysis. For it is usually assumed that the code can be described independently of the acts which it is used to explain. If this analytical and empirical separation of rule and action is not achieved, it is hardly possible to explain action in terms of compliance with a rule. Wieder points out, however, that this separation is seldom, if ever, attained, because the "social world of real events" which the sociologist attempts to explain has already been constructed by participants at least partly by means of the interpretative "application" of the code. In instances where participants have already used the code to categorize the actions in which the sociologist is interested, there will necessarily be an apparently direct relationship between rule and action. But it will be quite wrong to conclude that the actions, as interpreted, have been determined by the rule. To reason in this way is to lose sight of the interpretative work carried out by participants (or by the analysts themselves). The apparent fit between rule and action will have been constructed by participants; and, given the flexibility of rules, it could have been constructed otherwise. It should not, therefore, be taken for granted by the analyst.

If normative orders or codes are open-ended, flexible and continually reinterpreted as participants extend their meaning in the course of viewing new acts as instances of a code, "it is much more appropriate to think of the code as a continuous, ongoing process, rather than a set of stable elements of culture which endure [unchanged] through time."[15] (p. 161) If this view were to be taken seriously in the sociology of science, we could no longer remain content with having identified a group of "basic moral maxims." For, even if these maxims continued as part of the long term evaluative repertoire of science, we would expect that they would become hedged around with the kind of "supplementary" formulations observed by Zimmerman. At the very least, therefore, we would have to devote much more effort than we have done so far to observing the full range of formulations employed by scientists in connection with the recurrent activities of the scientific community. In addition, we would have to try to describe the kinds of interpretative procedures and resources employed by scientists in using evaluative codes, in a way similar to that begun by Wootton for therapeutic communities. Furthermore, we would have to try to show how scientists' use of normative formulations is linked to various kinds of interactional context and, perhaps, to participants' interests. Wieder[15] summarizes in the following words how his subjects made use of their code.

> The code operated as a device for stopping or changing the topic of a conversation. It was a device for legitimately declining a suggestion or order. It was a device for urging or defeating a proposed course of action. It was a device for accounting for why one should feel or act in the way that one did as an expectable, understandable, reasonable, and above all else acceptable way of acting or feeling. It was, therefore, a way of managing a course of conversation in such a way as to present the teller (or his colleague) as a reasonable, moral, and competent fellow. The code, then, is much more a method of moral persuasion and justification than it is a substantive account of an organized way of life. It is a way, or set of ways, of causing activities to be seen as morally, repetitively, and constrainedly organized.[15] (p. 158)

If Wieder is right, rules tend to be used as a fairly effective technique for closing down the process whereby the character of particular social events is negotiated.

> Thus, a resident's naming of a proposed act (such as the act of telling staff something or the act of participating in something) as a code-relevant event was a practically adequate answer to staff's requests, i.e. it effectively countered a request or demand in such a way that the resident was not required by staff to justify his refusal further. Moreover, on many occasions, staff not only "heard the code" and accepted it, but also acceded to residents' requests when the code was offered as grounds for action.[15] (p. 155)

From this perspective, then, analysis comes to focus, not on compliance with or deviation from a literally interpreted code, but on the ways in which participants use a code to establish acts as acceptable, rule-guided acts of a specific kind. Yet in the work of Wieder and Zimmerman there is one rather odd feature, namely, that acts are portrayed exclusively in terms of one single, coherent code, as if there were no alternative formulations available to participants. This is especially striking in Wieder's study because he confines his analysis to the ways in which the convict code is used by staff as well as residents to characterize and justify actions, even though he is observing the members of rehabilitation centers which have been established explicitly to combat the effects of criminal culture. It is surprising, therefore, that staff in particular appear never to use the formal morality of the penal system as an alternative evaluative repertoire. Wieder emphasizes that any particular act can be made to derive from various sets of rules which participants may see as incompatible in principle. This possibility is crucial to his argument. But in the course of his empirical analysis, he never mentions instances where different participants actually employed divergent codes or where the same participant used different codes in different settings. This may, of course, be due to the fact that the convict code was overwhelmingly dominant in the group studied by Wieder. I suspect, however, that in most groups and communities such alternative codes do exist, as Merton has suggested[1] and as Mitroff has sought to demonstrate in some detail for science.[6] If this is so, then the one-sidedness of Wieder's analysis may well be due to his concern with the longstanding "problem of social order." His central theoretical claim is that the characteristics of externality and constraint which have been attributed by sociologists in the Durkheimian tradition to moral codes, suicide rates, and other "social facts," as such, are actually constructed by participants in the course of social negotiation. "The interpersonal existence of social orders and their availability to perception and description is the achievement of the various methods entailed in an accounting-of-social-action." But do we have to suppose that participants are solely concerned to create social order and to maintain the existing state of affairs? Surely, in some circumstances, they will prefer to use rules and other cultural resources to depict others' acts as disorderly, improper and unacceptable; and, in so doing, *alter* may choose to employ another set of rules, instead of working within the code which *ego* is likely to adopt as providing the most acceptable version of his act. We will see in the next section that this is certainly so in science.

THE USE OF RULES IN SCIENCE

For the reasons outlined in the first part of this paper, sociologists have devoted little effort to documenting the rules actually used by scientists or to describing how these rules are employed in practice. What information we have on this topic has been produced largely as a byproduct of other intellectual endeavors. On the whole, sociologists have been content to re-state Merton's original brief formulations, as if they conveyed all that needed to be said about the norms of science. This is a tribute to the power of Merton's analysis, but it is hardly the best way of furthering our understanding of the social world of science. Consider, for example, the activities of communicating research results and assigning any ensuing rights of ownership. These potentially complex activities have been seen as regulated by one succinct rule, which as the following illustrations show, has been expressed with considerable uniformity and brevity by sociologists of science.

> Scientists should share their findings with other scientists freely and without favor.[16] (p. 79)
>
> Scientist should regard their discoveries as something they should share with others and not as their own private property.[11] (p. 15)
>
> Scientists should communicate their findings openly to other scientists. Keeping results secret is forbidden. Scientists cannot claim ownership of information, but only of recognition and esteem.[1] (pp. 464, 273–4)
>
> Scientists are not supposed to withold crucial aspects of their research beyond a reasonable period. . . After a research contribution has been published, the originator has no individual claims of ownership to the new idea, information or theory.[2] (p. 5)

There can be no doubt that the communication of results and the question of ownership are important issues for research scientists. Nor can there be much doubt that rules similar to the rule given above are often brought to bear on these issues by scientists. But in the light of the earlier discussion, we need to ask: Is this single formulation the only rule employed by scientists in this connection? And, on the assumption that there are at least a few "supplementary" rules, how are these rules applied and given meaning in particular instances?

One particular sequence of events which has been studied in some detail by Woolgar,[17,18] and which I have used before to try to throw some light on the complexities of rule-use in science,[8] is that associated with the discovery of pulsars. As both Zimmerman and Wieder note, rules tend to be formulated explicitly by participants when there is some kind of "trouble." The discovery of pulsars gave rise to trouble because a number of scientists claimed that those who made the discovery should have communicated their findings rather more quickly and more freely than they did; and also because some scientists claimed that credit for the discovery was wrongly allocated. From the traditional, normative view of science, the analyst would be inclined to try to ascertain whether there was in fact some deviation from "the norm of communality" in this case. Assuming that this could be established, then the reaction to certain aspects of the discovery on the part of scientists working on neighboring topics would be interpreted as the application of negative sanctions to a deviant act by those committed to the norm. However, this kind of interpretation becomes difficult to maintain, at least in this simple form, as soon as the analyst begins to take note of the variety of rules cited by participants as actually having guided their actions or as being the rules to which others should have conformed. The following list contains some of the rules for communicating results used by participants in the case of pulsars.[17–19]

1. Scientists should communicate freely and informally with all interested parties, even in the early stages of a discovery, because this will improve the quality of the scientific analysis.

2. Mention of current research outside one's own group should be kept to a minimum and, in the case of important discoveries, only those actively engaged in the research within the group should be allowed to know the details.

3. Findings of great significance should be released to as wide a circle of scientists as possible simultaneously. It will be necessary, therefore, to keep such findings secret until a formal article has been published in a journal of wide circulation.

4. It is improper that a scientist's immediate academic neighbors should learn through the journals that he has made a major discovery. It is necessary, therefore, to ensure that such colleagues are informed before formal publication.

5. It is legitimate to pass information to a favored group under special circumstances, for example, if this helps to improve the intellectual standing of an important national scientific organization.

6. It is normal scientific protocol that unpublished scientific information obtained informally should not be passed on to other groups.

7. Secrecy is offensive if it is selective, but acceptable if it is entirely uniform and if no exceptions are made.

8. Delay in communicating information which will allow others to take the research further, once the basic discovery has been published, is inexcusable.

9. Scientists can withhold (certain kinds of) data from publication if this prevents other groups from claiming some of the credit for the major discovery.

10. Scientists should not communicate important results, either formally or informally, until they are sure that there are no significant errors.

11. It does not matter if scientists withhold their results or not. Particularly when the findings are important, the delay will be short because scientists' self-interest will ensure that they publish quickly.

12. Researchers must be particularly careful not to release information in such a way that the first achievement of a graduate student is jeopardized.

13. Observers have a right to delay communication of their results in order to have time to attempt an interpretation before theorists step in and take over the more prestigious task of theoretical analysis.

14. Any results which might be of interest to the press must be kept as secret as possible.

15. Any discovery which raises the possibility of extra-terrestrial life is a special case and demands special treatment. If communication from other worlds were to be observed, a select council of leading scientists would have to be called, or some similar arrangement made, in place of formal publication.

There is a striking contrast between the simplicity and uniformity of sociologists' version of the norm of communality and the complexity and diversity of these formulations (and no attention has been given here to rules for allocating credit). It is clear from this one example that scientists have at their disposal a wide range of rules dealing with the communication of results, which can be used to account for and to claim legitimacy for a considerable variety of actions as well as to provide for many subtle distinctions between prescribed, allowable and unacceptable acts. The material presented above shows that we must investigate in much more detail than hitherto the actual formulations used by scientists before we can begin to understand how scientists' actions are related to rules.

In order to achieve this latter objective we must also try to explore how such a complex repertoire of rules is put into practice. We can begin with the observation that a particular scientist is often able to apply to one action two or more rules which seem to be literally incompatible (e.g. three and four above), without appearing to recognize any inconsistency. This confirms the claim made earlier that the meaning of such rules in relation to particular instances can never be taken for granted by the sociologist of science. The processes of interpretation which accompany the application of these rules are highly variable and they appear to depend partly on participants' subtle judgments about various aspects of the social world of science. For instance, interpretative judgments about the following features seem to be involved in the rules listed above: the importance of a discovery; other scientists' group membership and involvement in the research leading to the discovery; the audiences of various journals; the existence of groups suitable for "special treatment"; whether the discovery has special "nonscientific" implications; whether other researchers could pursue the scientific implications of the discovery more effectively; whether credit for the major discovery has been fully secured; how likely it is that all errors have been eliminated; whether any graduate student is likely to suffer from early disclosure; whether theorists are likely to skim off the cream; and whether the press would be actively interested.[7]

How, then, do participants link their repertoire of rules, each involving subtle interpretation of one or more of these features, to particular acts. The traditional sociological view has been that rules operate as determinate guides to action. But this conception remains convincing only as long as our ideas about participants' normative resources are kept unrealistically simple. If scientists were operating with one basic rule of communication which could be literally applied, then that rule could provide a clear guide for action. But the repertoire of rules and interpretative features identified above, incomplete as it undoubtedly is, is too complex to be used in this way. It is clearly impossible for any participant to run through the full range of conceivably relevant rules and their possible interpretations and permutations, before deciding how to communicate his findings. For the range of possibilities could be extended indefinitely. There is no point, therefore, in trying to ascertain whether a given act was guided by or performed in compliance with a particular cluster from the available rules. Indeed, to pursue this analytical objective is to misunderstand the nature of rules. It is to see the relationship between rules and action as causal rather than interpretative. It assumes a fixed connection instead of an open-ended link which participants can always reinterpret by bringing into play previously unused rules or by extending the meaning of those already cited. It seems, therefore, more profitable to ask: How do scientists use rules to characterize their own and others' acts; and how is use associated with variations in interpretative procedure and interactional context?

There is no space here for presentation of the empirical material necessary for any detailed answer to these questions. Consequently, I will do no more than offer a few brief comments, based on the case of the discovery of pulsars. It is clear that in this sequence of events rules were continually used as a resource for characterizing acts in ways that provided justification for them. For example, scientists who received informally data which had not been made generally available, described their decision not to pass this information to other groups as "in accord with established protocol." Similarly, they used the rule that information should be distributed to all interested parties as justifying their original request for the information. In characterizing their actions as conforming to a general rule which all scientists are expected to obey, researchers are portraying them as actions which, given the moral character of those involved, could hardly have been otherwise. At the same time, the morality of any person who fails to support their actions is put in question. In citing rule number one above, scientists could characterize the action of the discovery group in "withholding information" as a departure from the normal and proper state of affairs, while presenting their own request for information as a necessary rectification, as an attempt to reinstate a social order which was designed to foster the advance of scientific knowledge. Once they had obtained the information, however, rule number six could be cited as "preventing" them from passing it on, even though other groups could now claim it from them on the basis of rule number one.

This employment of rules by scientists seems similar to that reported by Wieder and Zimmerman, in the sense that rules are used to present actions as "morally, repetitively, and constrainedly organized." Conformity to selected rules is presented as obligatory; yet to the analyst it appears to be voluntary, variable and its form dependent on contingent interpretative work. In the case of pulsars, however, participants do not appear to have been limited to one coherent code which could be used as a routine means for closing off further debate. "Telling the code" was not, in this case, always or even usually an effective way of rebuffing others' claims for further justification. Characterization of an act in terms of any one rule typically led other scientists to challenge the propriety of the act thus characterized through the application of some other rule; or, in many instances, to question the adequacy of the initial interpretation of the original rule.

In the case of the discovery of pulsars, then, we appear to have a situation where participants did not succeed in using the available repertoire of interpretative resources to construct an agreed account of social reality. Each scientist or group of scientists proceeded to build up an interpretative gestalt of "what ought to have happened" as well as of "what really did happen." Participants used these gestalts to challenge particular acts through emphasizing the gap between actuality and propriety, the latter being defined by ostensibly literal readings of the relevent rules. It seems likely that in general, it is through accentuation of this supposed gap that moral condemnations are given their persuasive force. However, there is little evidence that these scientists had much success in changing each others' interpretations or in persuading others that they had implemented the wrong rules. For example, the original debate over secrecy flared into life again some six years after the discovery of pulsars, when this discovery was cited as one of the grounds for the award of a Nobel Prize. Indeed, on this later occasion, the original charges of secrecy were further embellished with assertions that credit for the discovery had been wrongly allocated. This subsequent exchange of views did not lead to any resolution of old differences, but rather to varied statements about how credit should be allocated and to extensive discussion of what constituted a discovery.[17] It is clear, therefore, that the negotiation of agreed versions of social reality is not always easily achieved in science and that participants' attempts to characterize acts either as rule-determined or as deviant are sometimes strongly resisted. Accordingly, it would be extremely useful if we could begin to devise analytical tools which helped to explain, not only how interpretative procedures work, but also how interpretative consensus and divergence are produced in different interactional contexts.

Concluding Remarks

There is no space here for me to develop these brief remarks further. However, I hope to have shown that there are good grounds for adopting a new approach to the study of norms in science. I have argued that scientists have a complex repertoire of rules which can be brought to bear on activities central to their professional life; that the relationship between these rules and scientists' actions is highly problematic and little understood; that the very character of the actions which sociologists seek to explain is in part defined by participants (and/or analysts) through the use of these rules; that scientists interpret and employ these rules in subtle and complex ways, making use of a variety of supplementary cultural resources and adapting the rules to the special characteristics of specific social situations; that there is no single, coherent code dominant in science, but rather a diverse variety of formulations which can easily be used by scientists to challenge any particular rule-based assertion; that, although rules are typically used by scientists as a means of claiming justification for their acts, such claims are frequently rejected; and, finally, that the open-ended nature of rules and the varied formulations available to scientists enables participants occupying different positions in a given social situation to use rules to create social disorder just as effectively as the social order which has interested most previous analysts of rule-use in other areas of social life.

Throughout this discussion I have concentrated on "moral rules." It should not be assumed, however, that the approach advocated here is to be confined to rules of this kind. This is nicely demonstrated in an analysis by Kuhn,[29] in which he shows that the procedural rules (or formal criteria) used by researchers in the course of establishing the validity of scientific theories are no more capable of providing determinate conclusions than the rules discussed above.

> When scientists must choose between competing theories, two men fully committed to the same list of criteria [rules] for choice may nevertheless reach different conclusions. Perhaps they interpret simplicity differently or have different convictions about the range of fields within which the consistency criterion must be met. Or perhaps they agree about these matters but differ about the relative weights to be accorded to these or to other criteria when several are deployed together. With respect to divergences of this sort, no set of choice criteria yet proposed is of any use ... one must go beyond the list of shared criteria to characteristics of the individuals who make the choice. One must, that is, deal with characteristics which vary from one scientist to another without thereby in the least jeopardizing their adherence to the canons that make science scientific. Though such canons do exist and should be discoverable (doubtless the criteria of choice with which I began are among them), they are not by themselves sufficient to determine the decisions of individual scientists. I am suggesting, of course, that the criteria of choice with which I began function not as rules, which determine choice, but as values, which influence it.[20] (pp. 324–5, 331).

If Kuhn's analysis is correct on this point, all rule-like formulations employed by scientists acquire their meaning through the interpretative work carried out by individuals in the course of social interaction in specific contexts. It follows that the interpretative approach I have advocated for the study of moral rules is identical with that required for the study of procedural rules and cognitive norms in science. The implications of this conclusion for the sociological study of science are very considerable.[21]

REFERENCES

1. MERTON, R. K. 1973. *The Sociology of Science.* The University of Chicago Press. Chicago and London.
2. GASTON, J. 1978. *The Reward System in British and American Science.* John Wiley & Sons. New York, N.Y.
3. BEN-DAVID, J. 1977. Organization, Social Control, and Cognitive Change in Science. *In: Culture and Its Creators: Essays in Honor of Edward Shils*: 224–265. The Univerity of Chicago Press. Chicago and London.
4. ZUCKERMAN, H. 1977. Deviant Behavior and Social Control in Science. *In: Deviance and Social Change.* E. Sagarin, Ed.: 87–138. Sage Publications. Beverley Hills and London.
5. COURNAND, A. & M. MEYER. 1976. The Scientist's Code. *Minerva* **14:** 79–96.
6. MITROFF, I. 1974. *The Subjective Side of Science.* Elsevier. Amsterdam and New York.
7. MARTIN, B. 1978. The Determinants of Scientific Behavior. *S.I.S. Review.* **2:** 112–118.
8. MULKAY, M. J. 1976. Norms and Ideology in Science. *Soc. Sci. Inform.* **15** (4/5): 637–656.
9. WITTGENSTEIN, L. 1953. *Philosophical Investigations.* Blackwell. Oxford.
10. MULKAY, M. J. 1969. Some Aspects of Cultural Growth in the Natural Sciences. *Social Res.* **36:** 22–52.
11. COTGROVE, S. & S. BOX. 1970. *Science, Industry and Society.* Allen and Unwin. London.
12. WOOTTON, A. J. 1977. Some Notes on the Organization of Talk in a Therapeutic Community. *Sociology.* **11** (2): 333–350
13. ZIMMERMAN, D. H. 1971. The Practicalities of Rule Use. *In: Understanding Everday Life.* J. D. Douglas, Ed.: 221–238. Routledge and Kegan Paul. London.
14. WIEDER, D. L. 1974. *Language and Social Reality.* Mouton. The Hague and Paris.

15. WIEDER, D. L. 1974. Telling the Code. *In: Ethnomethodology*. R. Turner, Ed.: 144–172. Penguin Books. Harmondsworth, Middlesex.
16. STORER, N. W. 1966. *The Social System of Science*. Holt, Rinehart & Winston. New York, N.Y.
17. WOOLGAR, S. W. 1978. The Emergence and Growth of Research Areas in Science, with Special Reference to Research on Pulsars. Unpublished Ph.D. Thesis. University of Cambridge.
18. WOOLGAR, S. W. 1976. Writing an Intellectual History of Scientific Development: The Use of Discovery Accounts. *Social Studies of Science*. **6** (3/4): 395–422.
19. EDGE, D. O. & M. J. MULKAY. 1976. *Astronomy Transformed*. John Wiley & Sons. New York, N.Y.
20. KUHN, T. S. 1977. *The Essential Tension*. The University of Chicago Press. Chicago and London.
21. MULKAY, M. J. 1979. *Science and the Sociology of Knowledge*. Allen and Unwin. London.

FISHER'S "REPRODUCTIVE VALUE" AS AN ECONOMIC SPECIMEN IN MERTON'S ZOO*

Paul A. Samuelson

Department of Economics
Massachusetts Institute of Technology
Cambridge, Massachusetts 02139

Robert K. Merton released from the closet the truth that scientists are human and work for the coin of applause from other scientists for their own innovations. Just as men will steal for bread and the ducats to buy bread, scholars will sometimes break the rules of the club in their pursuit of fame; or bend the rules; or put away temptation to do so, reluctantly if not firmly.

Nor is this surprising. Any stakes worth fighting for are worth being tempted about. I like to repeat my conversation with a dear colleague, H. A. Freeman. "Harold," I asked, "if the Devil came to you and in return for your immortal soul offered you a theorem, would you accept the Faustian bargain?" He replied without hesitation, "No." Then he went on to add, "But I would for an *inequality*."

Crimes are only in the eyes of the jury. I spoke of the rules of the club. The canons of scientists' etiquette were not ready-made on the seventh day of Genesis. If I knew in the sixteenth century the formula to solve a cubic equation, I wouldn't reveal it to you. I'd conquer examples with it for kudos and lucre. To protect my priority, I'd enunciate an anagram that conveyed nothing to the ignorant but proved to a Leibnitz or Wallace come-lately that my flag was already at the North Pole awaiting his arrival.

Clifford Truesdell claims that his hero Euler was the first scholar who tried in his writings to provide accurate accounts of just what his predecessors had done, citing earlier authors not merely for their errors, omissions and imperfections, but to inform the reader of the correct state of development of the subject under consideration. This suggests that present standards are but two centuries old, which is not long enough apparently for them to have penetrated all branches of discourse or reached all practitioners in any.

Crime for the most part really doesn't pay. The bank teller's lavish weekend has a price tag of hardtack for a decade. A fudged experiment or plagiarized lemma has a half-life of one academic semester. The talent needed to make crime pay can earn higher economic rent plowing the straight and narrow furrow.

To the sociologist and historian of science what is of interest is not the gross felony or even the indictable misdemeanor. Rather it is the study of the different styles of conduct open to the scientist who knows how to operate within the laws of the game. One is reminded of the fine art of tax *avoidance,* which is never to be mistaken for the crime of tax *evasion.*

People differ. A few are completely generous. They share their caviar with casual strangers, present their graduate students with crucial experiments and seminal hypotheses, and in the guise of anonymous referees reformulate whole articles. Again, Euler comes to mind. From the babbling of Maupertuis about teleology, Euler distilled the Principle of Least Action and presented it as Maupertuis' brain child. When he

*This work was supported in part by grants from the National Science Foundation 75-040530A01-SOC, National Institutes of Health 2-RO1-HD09081-03, and the Sloan Foundation.

126

0028-7113/80/0039-0126 $1.75/2 © 1980, NYAS

learned that the brilliant young Lagrange was about to publish researches on the calculus of variations, Euler held up his own deep memoir on the subject so as not to dim the light of the rising sun.

Euler could afford to be generous, you will say. He had an infinite treasure-house of good things. True. But John Bernoulli, perhaps Euler's most profound contemporary (although the claims of his brother James and son Daniel are comparably great), also had a treasure-house. Yet John proposed the famous brachistochrone problem in the calculus of variations primarily to try to show up his older brother James (he failed in this fratricidal ploy†); and John even pre-dated his unpublished writings in order to rob his own son of merited scientific priorities. If we dismiss John as pathological, we are given pause by the case of Gauss. No one was more generously endowed with genius and scholarly contributions than Gauss. Still the record is skimpy when it comes to generous acts on his part: when Wolfgang Bolyai, one of his few friends, reported that his son Johann had discovered and axiomatized non-Euclidian geometry, instead of congratulating the young genius for his genius, Gauss merely wrote back that his own unpublished works on the subject had followed the same track years before. Even if Johann Bolyai had not been depressed by this beyond the point of consummated suicide, Gauss's admirers will admit that truth is no defence for such a crime.

My interest on this occasion is neither with saints nor rascals. The former are all too few. The latter are born all too plentiful. Thus, Laplace was unscrupulous to a fault in appropriating the work of others and sufficiently absent-minded to plagiarize his own earlier works. (Yet before we throw stones, we are given pause by the fact that Laplace was uniquely able to add value to his purloined rubies and to recognize their intrinsic worth.) In my field of economics, the great Swedish scholar of the World War I epoch, Gustav Cassel, was notorious for never acknowledging the sources of his contributions. "Ten per cent Leon Walras and 90 per cent water," is how my old teacher Schumpeter used to dismiss Cassel. Cassel's secretary, in her biography of him, attributed his bad scholarly manners to Cassel's hatred of his father, which is supposed to have blocked from his memory all he learned in his earlier decades (and, apparently, to have fastened that habit on him for life!).

FOIBLES OF GENIUS

Here I want to concentrate on the pattern of men who are so highly original that they find it more economical to work out a thing themselves, even if in doing so they are repeating without acknowledgment what is already in the literature—easier to forge a ballbearing than find one in inventory. Albert Einstein was such a scholar. So, I believe, was Enrico Fermi. R. A. Fisher, who revolutionized both mathematical statistics and neo-Darwinian genetics, provides me with a fine example for Mertonian dissection.

In my own field of economics, Columbia's John Bates Clark never read anything.

†The Mertonian jury of historians of science can never be permanently discharged; its verdicts are always tentative and subject to change. The conventional view, by G.A. Bliss and others, is that James Bernoulli is really the victor in the brachistochrone duel with John: although John has the more brilliant and glib solution through his analogy with the way geodesics of light bend in a medium of variable refraction index, James' more clumsy variational technique in fact established the firm foundations upon which later writers built the calculus of variations. Imagine then my surprise to learn from the speech of the great Carathéodory at Harvard's 1936 Tercentenary that John's local analysis may be deemed the first instance of the modern deeper approach.

Not only was he the first to discover the marginal productivity model of income distribution (an accomplishment that makes me proud to wear his medal). Clark also rediscovered, a decade late and in defective form, the theory of marginal utility. His son, John Maurice Clark, also a famous scholar, also read nothing. "We used to have to put a book in front of Maurice," Frank Knight once told me, "and open the book to the proper page, to get him acquainted." Perhaps this was just as well. J. M. Clark's most famous finding, his 1917 enunciation of the Acceleration Principle—relating the *level* of investment spending to the *rate of change* of consumption (or total final) spending—might never have occurred if he had known of Bickerdike's 1914 anticipation, or Hawtrey's 1913 approach, or Aftalion's . . .

Einstein himself didn't know whether or not he had heard of the Michelson-Morley experiment on the equal velocity of light in all directions when he axiomatized his special theory of relativity. Whittaker, who had a grudge against Einstein, reproached him for his cavalier ignoring of Poincaré's contemporaneous work: what Whittaker failed to mention was that there was nothing personal about this, Einstein rarely pauses to recognize property rights of any earlier scholars. When Fermi wrote up his lecture notes on thermodynamics, he assigned an assistant the task of spiking them with footnote references. When someone would come to Richard Feynman with an approximation result in physics, Feynman would often take down a notebook from his shelf and verify from his unpublished version of the exact solution that his visitor had indeed correctly identified the first term in the complete expansion. Men have been shot for less.

My own titanic colleague, Sir John Hicks of Oxford, always tells us how he came to his ideas. "I could have kicked myself," his prefaces relate about a delayed perception. "Already, I had perceived the broad outlines of my theory . . ." What his footnotes rarely contain is informed and precise appraisal of past and present parallel researches. Sometimes this results in what Gunnar Myrdal years ago called "attractive but unnecessary Anglo-Saxon originality." Sometimes, as in the rediscovery of Markowitz's well-publicized portfolio model of efficient mean-variance tradeoff, it results in anti-climax. Redundancy, I am prepared to argue, is only a minor blemish: it causes a few trees to be unnecessarily axed, the spilling of some rather cheap ink, a waste of readers' limited time and attention. But, before you go hacking with Ockham's razor, be sure you don't cut muscle along with fat. Because the sun does not rise until a scholar of Hicks' originality opens his eyes in the morning, we get a fresh view of any subject he addresses. Often I find novel Hicksian insights in the treatment of old topics. (I once explained to Joan Robinson, after she remarked that English economists wouldn't mind reading American economics if only they knew where to begin on that vast continent, "We New World economists take unfair advantage: we can read both English and American.")

R. A. FISHER, A CASE IN POINT

It is Sir Ronald Fisher—R. A. Fisher to my generation of statisticians—that I am concerned with here. Fisher was a genius—if that term has a meaning. He founded, and revolutionized, the subject of statistics. For the quarter century after 1912, statistics *was* R. A. Fisher. Great as Fisher was in the domain of mathematical statistics, one is told he was equally great in the domain of genetics. It is regarded as a reproach that the Nobel Prize in biology was never awarded to R. A. Fisher, J. B. S. Haldane, and Sewall Wright. Fisher's *The Genetical Theory of Natural Selection*[1] is a classic, regarded with the awe to be accorded a bible and containing some of the mysteries and obscurities of that sacred work.

Here I shall deal with one small gem from this book, Fisher's novel and ingenious concept of "reproductive value." He bequeathed us scores of such jewels, many of which might serve my purpose as well were it not for the accident that, in training for understanding Fisher's nearby Fundamental Theorem of Natural Selection, my attention was fastened upon the ambiguities implicit in Fisher's reproductive-value concept and his exposition of it, creating in me an itch that could not be scratched out until I found myself writing four (brief, unrefereed) articles on the subject.

Since some of my words might seem less than fulsome, I ought to begin with a deposition of R.A. Fisher's greatness. Like the great John Maynard Keynes, Fisher probably increased the impact of his lasting discoveries and novelties by the polemical mode of his scientific writing. Had Fisher been able to narrate the steps in reasoning by which he came to his theorems, they might have lost a sense of magic that invests them. Each reader might then be able to say, "Shucks, there with the grace of God might go I, and what's so remarkable about anything as freely got as that?" Obscurity is neither a necessary nor a sufficient condition for magnifying one's impact; but, conjoined with a message worth sending, it can bull up market price. The labor theory of value applies in the history of ideas. After you have hunted through 2 million words in *Das Kapital,* you esteem the catch. The wood I chop warms me twice. Newton in geometric Latin is more profound than in the freshman textbook. Unheard melodies are sweeter yet.

I wrote had Fisher "been able to," rather than "willing to," because there is a question as to whether his was not the intuitive mind that arrives at a gestalt without being able to rationalize its historical steps. My old Harvard teacher, Edwin Bidwell Wilson ("E.B."), who was a good acquaintance of Ronald Fisher and a genial admirer and observer, used to comment on Fisher's poor eyesight. "Perhaps that is why Fisher was so good at the n-dimensional geometry by which he derived his 'F' and other notable distributions," E.B. used to say. "Certainly it explains the extraordinary number of printing errors that permeate his fundamental memoirs." Or, to take another example of Fisher's remarkable flights of intuition, Sir Maurice Kendall has commented somewhere on the mystery of Fisher's mode of discovering the intricate k-statistics so useful when sampling moments, semi-invariants, and cumulants of theoretical probability universes. Not only could Kendall not reconstruct the paths of reasoning, but Fisher could also not do so on interrogation.

Was it Virginia Woolf (or Stephen) who said, "How do I know what I think until I hear myself saying it?" The quip that Hegel didn't really understand his own philosophy until he read the French translation is a minor truth next to that which claims that an idea does not really exist until it can be plagiarized. We have all had a brilliant Coleridgian colleague whose head is full of deep perceptions that can't quite be grasped. How do chimeras parse *cogito ergo sum?* My point is that, after perusal of Fisher's texts and reflection on them, I am left with the hypothesis that not infrequently his unclarities are not perverse thumbnosings at the reader, not fudgings over the fine details of the syllogism or its limitations, but rather they faithfully mirror unclarities in his own thought. Why, in connection with matters that he wrote on over a period of more than a quarter of a century, he did not choose to devote the time to think them through, one can only speculate about.

Cassel, his secretary exonerated on the grounds of friction with his father. Fisher's rudeness has been excused on the grounds that he was as much sinned against as sinning. My generation of students were like children observing their elders at a drunken party, so titillated were we at the display of fireworks by Karl Pearson and Ronald Fisher. Was there ever another generation of savants with similar scholarly manners? It used to be said that Planck and Einstein were such nice people that they elevated a whole generation of German physicists (no mean act of levitation).

Example works both ways. When I expressed to E.B. my shock at the motion of thanks emitted by Fisher after an invited 1935 speech by Jerzy Neyman before the *Royal Statistical Society,*[2] it was explained to me that Karl Pearson had treated Fisher so badly as to make this understandable. And Alfred North Whitehead once boasted at Harvard's Society of Fellows that he had been the only person in Britain who managed to stay on good terms with both the Mendelian advocate, William Bateson, and the Galtonian advocate, Karl Pearson. Each, he said, was impossible to use as a referee: the slightest commendation in a manuscript of work from the enemy camp was enough to outweigh any excellence it possessed.

Science, some sage said, advances funeral by funeral. Fisher was out of tune with his times after 1940. Statistical decision making he denounced as a fascistic tool. Confidence intervals did not gain his confidence. Bayesian tools he eschewed; but after decades of declaiming against equi-probability of the unknown, Fisher explained that his fiducial probability involved a proper assumption of balanced ignorance. Neyman-Pearson rejection regions based on minimizing Type B errors were rejected by Fisher as the figment of those who did not understand practical matters—and this from the man who chided Gregor Mendel for reporting ratios *too near* to 3-to-1, and who casually releases in *Statistical Methods for Research Workers* the bombshell that *too-small* values of chi-square are as bad as the *large* ones upon which his tables are built!

Genius can get away with methods unforgivable in talent? That is the reverse of truth, precisely because of Merton's syndrome of the Matthew Effect.[3] To him who hath shall be given—and given also the power to take away. Because Fisher's work in design of agricultural experiments was so good, his unfortunate work on smoking and longevity proved so very unfortunate in misleading statisticians of his claque. Because his analysis of genetic models was so pathbreaking, Fisher's final chapter, in which he seriously purports to deduce the decay of civilization from the propensity of heritably unfecund heiresses to capture as mates and render insterile the most successful (and genetically able) elements of the population, was taken with comical seriousness. My point is that it takes the wisdom of Maxwell's Demon to separate out the good from the bad in the *oeuvres* of an acknowledged genius—the more reason for refusing him license.‡

FISHER'S VERSION OF THE LOTKA MODEL

My purpose is to describe, and comment on, two distinct components of Fisher's 1930 exposition: his rendering without citations of A.J. Lotka's model of population that is self-propelled by invariant age-specific mortality and fertility schedules; Fisher's own discovery and cryptic explication of the novel concept of "reproductive value" of each age-group.

Here, in selected words of Fisher, is the first of these.[4]

> ... a life table ... shows for each year of age, of the population considered, the proportion of persons born alive who live to that age ... ℓ_x [being] the relative number living to age x, the death rate at age x is given by: $\mu_x = -\mathrm{d} \log_e(\ell_x)/\mathrm{d}x$.
>
> The life table concerns itself only with death and not at all with reproduction. To repair this deficiency it is necessary to introduce a second table giving rates of reproduction in a manner analogous to the rates of death at each age. Just as a person alive at the beginning of

---

‡A. E. Housman was asked about an item that he refused to reprint in a collection of his classic writings, whether he did not think it good. He replied: "It is good, but not good enough for me."

any infinitesimal age interval dx has a chance of dying within that interval measured by $\mu_x dx$, so the chance of reproducing within this interval will be represented by $b_x dx$, in which b_x may be called the rate of reproduction at age x. Again, just as the chance of a person chosen at birth dying within a specified interval of age dx is $\ell_x \mu_x dx$, so the chance of such a person living to reproduce in that interval will be $\ell_x b_x dx$. . . . If we combine the two tables . . . of death and reproduction, we may . . . calculate the expectation of offspring of the newly born child . . . $\int_0^\infty \ell_x b_x dx$ [which] determines whether . . . reproductive rates are more or less than sufficient to balance the existing death rates. If its value is less than unity . . . [the] population . . . would . . . certainly ultimately decline in numbers at a calculable rate. Equally, if it is greater than unity, the population biologically speaking is more than holding its own, although the actual number of heads to be counted may be temporarily increasing . . . . corresponding to any system of rates of death and reproduction, there is only one possible [stable age distribution] . . . proportional to $e^{-mx}\ell_x dx$ . . . [where m is the one and only existent real solution to] $\int_0^\infty e^{-mx}\ell_x b_x dx$ . . . equated to unity. . . . The number m, is thus implicit in any given system of rates of death and reproduction, and measures the relative rate of increase or decrease of a population when in the steady state [and] . . . m may appropriately be termed the Malthusian parameter . . .

Those familiar with A. J. Lotka's post-1911 researches on a population that is self-propelled toward stable exponential growth by invariant age-specific mortality and fertility parameters will recognize in Fisher's above account a useful description of the Lotka model. There is, however, no mention of Lotka's name and no reference to work of other earlier writers such as R. R. Kuczynski, who had independently and as a disciple of Böckh published on the related concept of *net reproduction rate*.[5]

Inasmuch as Fisher was an undergraduate in 1911 when Sharpe and Lotka[6] appeared, and since he was then presumably more interested in mathematics and astronomy than in the life sciences, one can doubt that Fisher knew this earliest work—or that of Lotka[7,8] (1913, 1922). Indeed a person of Fisher's originality may well have worked out the whole theory by himself in ignorance of its prior existence—a doubleton that Robert Merton would not be surprised at. On the other hand, Lotka's 1925 classic, *The Elements of Physical Biology*,[9] achieved wide circulation and might have been expected to pique Fisher's curiosity. Its Equation (33), p. 118, clearly presents the relation that Fisher invoked to determine the Malthusian parameter m; several of Lotka's remarks on selective slaughtering of the old rather than young do smack of the "reproductive value" concept; and Fisher's p. 29 remark that the human death rate is at its minimum at age 12 could well have come from Lotka's p. 151 observation rather than from the more obscure original Raymond Pearl source.

All this presents the Mertonian detective with a nice problem in maximum likelihood or Bayesian estimation. What are the subjective odds that, if a definitive document is later found that *settles* whether Fisher did or did not know about the Lotka system, we as informed scholars will bet that it will reveal Fisher's discovery not to have been independent? To formulate these odds, one naturally turns to other discoveries of Fisher. Fisher[10] presents without citation of von Neumann's work on game theory what is clearly a formulation of the mixed-strategy saddlepoint solution to a 2-person zero-sum game (that of "her" discussed in Todhunter's standard history of probability). My considered guess is that Fisher did not then know of von Neumann's 1928 German mathematical article on game theory[11] (but he might have vaguely heard of Borel's less complete speculations?). When as an undergraduate Fisher proposed his 1912 maximum-likelihood criterion for estimation, one can doubt that he knew of similar work a century earlier by Gauss (or, for that matter, but here we are on trickier ground, of the more recent work in Edgeworth (1908) that contained at least implicitly the notion of "efficiency" of the maximum-likelihod estimator). The written remark of H. L. Seal[12] (1967) is relevant to our forming subjective probability judgements: ". . . we have heard it stated that Fisher in private conversation showed himself familiar with the work of the earlier mathematicians in Europe."

Seal adds what I can only agree with: "It is no discredit to the genius whose published output is large and continuous that he failed to take the time to review all previous literature." I should also add that, personally, it provides an even greater challenge to me to try to add to what my predecessors have done, *within the constraint* of formally pointing out their accomplishments: just how high is the altitude reached by the shoulders of the giants that bear me is the necessary base from which I can estimate my pygmy reach. And one should note that, when Fisher wished to mete out faint praise to Karl Pearson, he was capable of quoting the little known 1872 work of Helmert on the derivation of what came to be called Pearson's chi-square distribution.

We do not have to go to the works of an Einstein or Fisher to discern a not uncommon gambit, part of Stephen Potter's one-upmanship as played by scholars, in which a safely-dead scholar is used to eclipse a live rival. If I am to be anticipated, pray that it be by someone centuries back! To call m the Malthusian parameter, rather than say "the Lotka intrinsic rate of natural increase" may not have been such an unconscious ploy, but we cannot rule out that it was.

Undoubted Originality

I move on now to the part of Fisher's discussion that is unquestionably novel—his defining of "reproductive value of each age group" in terms of an almost mystical analogy to the economics of compound interest. Again his own words can be adequately sampled.[13]

> In view of the close analogy between the growth of a population supposed to follow the law of geometric increase, and the growth of capital invested at compound interest, it is worth noting that if we regard the birth of a child as the loaning to him of a life, and the birth of his offspring as a subsequent repayment of the debt, the method by which m is calculated shows that it is equivalent to answering the question—At what rate of interest are the repayments the just equivalent of the loan? . . . the Malthusian parameter of population increase is the rate of interest at which the present value of the births of offspring to be expected is equal to unity at the date of birth of their parent [$1 = \int_0^\infty e^{-mx}\ell_x b_x \mathrm{d}x$].
>
> . . . The analogy with money does, however, make clear the argument for another simple application of the combined death and reproduction rates. We may ask, not only about the newly born, but about persons of any chosen age, what is the present value of their future offspring; and if the present value is calculated at the rate determined as before, the question has the definite meaning—To what extent will persons of this age, on the average, contribute to the ancestry of future generations? The question is one of some interest, since the direct action of Natural Selection must be proportional to this contribution. . . . by the analogy of compound interest the present value of the future offspring of persons aged x is easily seen to be given by the equation
>
> $$v_x/v_0 = \frac{e^{mx}}{\ell_x} \int_x^\infty e^{-mt}\ell_t b_t \, \mathrm{d}t.$$
>
> Each age group may in this way be assigned its appropriate reproductive value . . .
>
> A property that well illustrates the significance of the method of valuation, by which, instead of counting all individuals as of equal value in respect of future population, persons of each age are assigned an appropriate values v_x, is that, whatever may be the age constitution of a population, its total reproductive value will increase or decrease according to the correct Malthusian rate m, where as counting all heads as equal this is only true in the theoretical case in which the population is in its steady state.

Mere commonsense says: Choosing to save the child does more for the future population than to save the grandparent; if the child may die before puberty, saving the adolescent is more economical still. Fisher advances from this commonplace to an exact quantitative evaluation of the proper tradeoff between persons of different ages when it comes to determining the level at which the population will eventually double and redouble.

Moreover, as always, when we explain commonsense we go beyond it to the rare atmosphere of new truth. In a declining population—like ours in which few children die but lack brothers and sisters—the new babe has more reproductive value than the nubile maid. Why? That is what Fisher's formula spits out when you insert a negative interest rate into its slot. "But *why*?" you insist on asking. Perhaps for the reason the talmudists agreed the moon is more important than the sun: true the moon gives less light than the sun, but it does so during night's darkness when we need it most. The babe's gift will come later and will have melted away less by the time it arrives.

What is most awesome about Fisher's exposition is its oddness. A mysterious if not mystical analogy is made to economics. Just as a bond's yield is that rate of interest which will discount its future cash payments to its present market price, Fisher says a population's growth rate is that interest rate which will discount all a new babe's future births to the unit life bestowed on her. Who enters into this contract? Enforces it? Perceives it? Why, at each age of life, should the present discounted-value of her future expected offspring, calculated with the growth rate that the population does not *yet* experience (but will ultimately experience) be a proper measure of what each initial member of that age group will contribute to the ultimate level of a population's growing?

As Professor Keyfitz has mentioned in our private conversation, once the problem is well-posed of determining how the initial age distribution determines the coefficient of the dominating exponential in the model's exact solution, the solution in terms of a certain characteristic row vector or eigenfunction is complicated but not mysterious. What is amazing is how Fisher apparently bypassed the straightforward and leaped to the solution immediately by means of the compound-interest analogy. As a demographer and sociologist, Professor Keyfitz may be excused for not perceiving intuitively the economic interest approach. But I am an experienced economist, one who has made his profits so to speak from the study of compound interest and capital theory, and yet I must confess—nay insist—that the argument is a mysterious one and even (looked at from the standpoint of a general, not necessarily linear, model) an inconclusive one. If someone told me that Fisher had actually found his solution in the straightforward way and then—after the fact!—puzzled out how to express it in compound interest terms, I would not find that hard to believe. And yet there is no evidence for such a scenario and no warrant for asserting its validity.

It occurred to me to investigate what Fisher's biologist readers have made of his compound interest analogy. As might have been expected, they seem simply to take it on faith, reiterating its assertions. As a representative specimen, E. O. Wilson's magisterial *Sociobiology*[14] may be cited. In the course of its excellent, brief description of reproductive value, the relevance of the Malthusian parameter as a discount factor is not only not really explained, but is in fact omitted from some of the key literary explanations.

MATHEMATICAL PROOF

Before turning to my positive analysis of the economic valuations definable by demographic or biological models, let me give a brief account of the straightforward approach to reproductive value that a mathematician would take. Write $B(t)$ for (female) births at time t; $N(x, t)$ for the number (of females) of age x at time t (i.e., the "density" of females between x and $x + dx$), where $N(0, t) \equiv B(t)$; ℓ_x for the fraction of new-born (female) babies who survive to age x; and b_x for the rate at which women of age x bear female babies. Then Lotka's model—and Fisher's!—is definable by

$$B(t) = \int_0^\beta \ell_x b_x B(t - x) \, dx + \int_0^\beta (\ell_{x+t}/\ell_x) \, b_{x+t} N(x, 0) \, dx$$

$$= \int_0^\beta \ell_x b_x B(t - x) \, dx + G(t), \qquad 0 < \alpha < \beta < \infty \tag{1}$$

$$0 \le G(t); G(t) \equiv 0, t > \beta; \ell_x b_x > 0 \quad \text{for} \quad \alpha < x < \beta$$

The Volterra integral equation of the second kind in (1) is known to have an explicit solution of the form

$$B(t) = G(t) + \int_0^t \phi^*(t - y) G(y) \, dy$$

$$= 0 + \int_0^\beta \phi^*(t - y) \left[\int_0^\beta \ell_{x+y}/\ell_x) b_{x+y} N(x, 0) \, dx \right] dy, t > \beta \tag{2}$$

$$= \int_0^\beta N(x, 0) \left[\int_0^\beta \phi^*(t - y) \, \ell_{x+y/\ell_x}) b_{x+y} \, dy \right] dx, \quad t > \beta$$

$$= \int_0^\beta N(x, 0) v(x, t) \, dx, t > \beta \tag{3}$$

$$= 0 + \int_0^\beta v(x, t) \, N(x, 0) \, dx, \text{t} > \beta$$

where $\phi^*(t - y)$ is the resolvent kernel to the kernel $\ell_{t-y} b_{t-y}$ (as discussed in Samuelson[15]) and where $v(x, t)$ is the generalization *to any* t for Fisher's reproductive value of an initial age element. By convention, $\phi^*(t)$ vanishes for negative t.

By definition, $v(x,t)$ represents the contribution to births t periods later of an initial female of age x:

$$v(x, t) = \int_0^\beta (\ell_z/\ell_x) b_z \phi^*(t - z + x) \, dz \tag{4}$$

In the limit as $t \to \infty$, $\phi^*(t)$ becomes essentially exponential, ce^{mt}, and so

$$v(x)/v(0) = \lim_{t \to \infty} [v(x, t)/v(0, t)] = \int_x^\beta (\ell_z/\ell_x) b_z \exp[m(x - z)] \, dz \tag{5}$$

in agreement with Fisher's formula.

The resolvent kernel $\phi^*(t)$ has the clear interpretation as the stream of subsequent female births occasioned t periods from now by a unit pulse of female babies now. Just as the inverse of a Leslie matrix can achieve the solution of a problem with any initial conditions, so the resolvent (or "inverse") kernel can do the same in the continuous case. The sole integral equation for $\phi^*(t)$ is

$$\phi^*(t) = \ell_t b_t + \int_0^t \ell_{t-z} b_{t-z} \phi^*(z) \, dz \tag{6}$$

$$= \ell_t b_t + \int_0^t \ell_{t-z} b_{t-z} \ell_z b_z \, dz \tag{7}$$

$$+ \int_0^\beta \ell_{t-x} b_{t-x} \left[\int_0^\beta \ell_{t-z} b_{x-z} \ell_z b_z \, dz \right] dx + \cdots$$

where (7) is a uniformly converging iteration series.

An interesting approach to $\phi^*(t)$ comes from pretending that $\ell_x b_x e^{-mx}$ is the death-rate function for some associated process. Such a death-rate function defines a renewal (or, better, "replacement") function, $R(t)$, by the standard renewal integral equation

$$R(t) = \ell_t b_t e^{-mt} + \int_0^t \ell_{t-z} b_{t-z} e^{-mt+mz} R(z)\, dz \tag{8}$$

$$\lim_{t\to\infty} R(t) = \left[\int_0^\infty \ell_x b_x e^{-mx}\, dx \right]^{-1} \tag{9}$$

$$\phi^*(t) = R(t)\, e^{mt} \tag{10}$$

$$\approx \left[\int_{\infty}^0 x\ell_x b_x e^{-mx}\, dx \right]^{-1} e^{mt}, \quad t \gg 1$$

The equivalence of (8) and (6) is seen after (9) is substituted into (6) and (8) results.

Note that this straightforward proof does not utilize either Laplace transforms or the Hertz-Herglotz infinite exponential expansions of Sharpe and Lotka. So long as b_x vanishes after some finite age, there is no lack of rigor in the second of these approaches, as has been shown by Lopez.[16] Using it, we find that (1)'s solution can be written as

$$B(t) = Q_0 e^{mt} + \sum_{j=\pm 1}^{\pm\infty} Q_j e^{m_j t} \tag{11}$$

$$= Q_0 e^{mt}[1 + \epsilon(t)], \quad \lim_{t\to\infty} \epsilon(t) = 0$$

where

$$Q_0 = \frac{\int_0^\beta N(x,0) \int_x^\beta (\ell_z/\ell_x)\, b_z \exp[(x-z)m]\, dz}{\int_0^\beta z\ell_z b_z e^{-mz}\, dz} \tag{12}$$

and the (m_j, m_{-j}) are conjugate complex roots with coefficient of their real part less than m.

The reader may be referred to Keyfitz[17] for the best treatment known to me of Fisher's reproductive value as applied to the Lotka-Leslie one-sex linear model, the one case for which it is exact.

Fisher writes as though his reproductive value formula would apply—no doubt with various straightforward but minor qualifications—to realistic biparental (i.e. bisexual) models involving both fathers and mothers. He seems to believe that the sexes are tautologically equal in total reproductive value (in that every baby has exactly 1 mother and 1 father), and hence either can stand for half the whole. Fisher also recognizes that a mere increase in numbers will cause the effective environment to deteriorate and make his ℓ_x and b_x functions be also functions of total numbers, such as $\int_0^\infty N(x,t)dx$. This destroys the strict linearity of his model and vitiates the principle of superposition on which his intuitive syllogisms depend. Also, but this is presumably not substantively a strong effect, Fisher recognizes that females beyond the menopausal climacteric of age β may add to the future population by the care they provide for their kin; again, such plausible effects cannot be correctly handled by making $b_{\beta+x}$ positive instead of zero. Since, however, Samuelson[18-21] has gone into these aspects at sufficient length, my concluding section on the economics of reproductive valuations will touch upon them only in connection with the demonstration of lack of *logical* connection between various of Fisher's propositions.

ECONOMICS OF BIOLOGICAL STATE

I conclude by investigating more deeply the analogy between economists' price and Mother Nature's scarcities.

1. It is standard to calculate the present-discounted-value of persons of different ages. This is done for juries in damage awards, for clients of life insurance to help determine their coverage needs, for scholars studying "human capital" and the economics of education. [See Dublin and Lotka[22], Walsh[23].] However, it is usually *not* to the b_z fertility function that the (ℓ_z/ℓ_x) conditional survival fractions and $\exp[r(x - z)]$ discount factor are applied, but rather to some measure of wage earnings by age.§ And the interest rate used, r, is as Fisher noted generally not of the same magnitude as the Malthus-Lotka m. [See, however, Samuelson[25] for a "biological" interest rate; and Phelps[26] for the concept of the golden-rule interest rate, equal to the economy's natural growth rate of labor (in biological or efficiency-unit terms) and which leads to production methods that maximize consumption per head.]

2. Purely by the definition of present-discounted-values and without reference to any biological relations as such, any repeated investment project whose initial cost is precisely equal to what is needed to make it just pay will at every instant of time earn on the portfolio of accumulated and reinvested returns the postulated market interest rate. Fisher's demonstration that the calculated total reproductive value, $\int v(x)N(x,t)dt$, grows *from the beginning* at his postulated rate m is a brilliant instance of this general truth. (Example: You buy a bond or truck that gives you at stated intervals cash payments and then a final payment. Its initial cost to you is set at the present-discounted-value of such an income stream; and each cash payment you plow back into similar new purchases, never withdrawing any net cash from your total portfolio of such investments. Then, as shown in Samuelson[27], your principal grows at each instant at the constant interest rate postulated. What Fisher does not touch upon is the still more general tautology: even if the interest rates are *not* constant over time, being $r(t)$ rather than r, it will still be true that your portfolio grows at each t at the exact percentage rate $r(t)$ provided that all costs of investment projects are at their true present-discounted-values.

3. Now I examine processes that generate their own growth: 2 rabbits give you 4 every period; or a Lotka-Leslie model postulates that female babies produce so many female adults who produce so many female babies at different ages. If we reckon the interest rate in terms of money, and if the r is not equal to the Malthus-Neumann rabbit "own rate of interest," it would violate the rules of competitive arbitrage for the money price of a rabbit to be constant over time. If the rabbit interest rate is 20 percent per annum and the money interest rate is 5 percent, then the money price of rabbits must fall in proportion to $(1.05/1.20)^t$.

In the rabbit case, so long as there are no overlapping generations, the population and births will grow exponentially from the beginning at the $(1.20)^t$ rate. No reproductive value weighting is needed, and 1 rabbit does equal the present-discounted-value of its 1.2 offspring at the rabbit-own-rate-of-interest of $m = .2$.

§A partial exception occurs when Fogel and Engerman[24] seek to explore the economics of ante bellum slavery in those regions and sectors where slaves are bred to breed slaves. Then, ignoring byproducts such as slaves' useful non-breeding services, they tend to find that nubile females are worth more than post-menopausal elders or newborn female babies. However, the proper present discounted value computations are done with the interest rate earnable on comparable investments and not on some biological Malthusian rate. Furthermore, the imputed value of land and other services needed to cooperate with people in keeping them fit and able to reproduce babies must be reckoned in, by methods obvious to economists but much more complex than Fisher needed for the idealized Leslie model.

For the Lotka-Leslie case in which females are mothers at various ages of life, there is defined by the process what Boulding[28] would call "the internal [own] rate of return of interest," which is precisely the Lotka-Fisher rate m that discounts the sum of all future female daughters into equality with unity.

Thus, the only female-baby own-rate-of-interest that could be constant over time in a perfect competitive market is the Malthus-Fisher m of the Lotka-Leslie process.

4. The Lotka process, particularly in its Leslie[29] version involving discrete time units and age intervals, is a special and simple instance of the von Neumann[30] self-generating general-equilibrium model, in which any good can be producible as output out of inputs that are the system's output one period earlier. Specifically here,

$$b_x \text{ female babies and } p_x \text{ females of age } x + 1$$
$$\text{are produced by} \tag{13}$$
$$1 \text{ female of age } x$$

Here $[b_x] = [b_1, b_2, \ldots, b_\beta, b_{\beta+1} = 0]$ with $b_\beta > 0$, $b_{\beta-1} > 0, \ldots$. Also $[p_x] = [p_1, p_2, \ldots, p_{\beta-1}, 0, 0]$, with $0 < p_j = \ell_{j+1}/\ell_j \le 1$.

Then competitive steady-state price equilibrium requires of the interest rate r and the respective prices of a female of age x, $[\pi_x] = [\pi_1, \pi_2, \ldots, \pi_\beta, \pi_{\beta+1}]$, that the following cost-of-production relations obtain

$$\pi_1 b_x + \pi_{x+1} p_x = \pi_x (1 + r), (x = 1, 2, \ldots, \beta) \tag{14}$$

This says that I must earn the common interest rate on my outlay in buying 1 female of age x as input when she produces for me b_x daughters and turns into p_x females of age $x + 1$ and I can sell these joint products at their respective prices.

In matrix terms we have, where bold face type denotes a matrix or vector;

$$\mathbf{N(x, t + 1)} = \begin{bmatrix} N(1, t + 1) \\ \vdots \\ N(\beta + 1, t + 1) \end{bmatrix} = \begin{bmatrix} b_1 & b_2 & \cdots & b_\beta & 0 \\ p_1 & 0 & \cdots & 0 & 0 \\ 0 & p_2 & \cdots & 0 & 0 \\ \cdot & \cdot & \cdots & \cdot & \cdot \\ \cdot & \cdot & \cdots & \cdot & \cdot \\ 0 & 0 & \cdots & p_{\beta-1} & 0 \\ 0 & 0 & \cdots & 0 & 0 \end{bmatrix} \begin{bmatrix} N(1, t) \\ \vdots \\ N(\beta + 1, t) \end{bmatrix} \tag{15}$$

$$= \mathbf{LN(x, t)}$$

$$\boldsymbol{\pi} \mathbf{L} = [\pi_1, \ldots, \pi_\beta, \pi_{\beta+1}] \mathbf{L} = [\pi_1, \ldots, \pi_\beta, \pi_{\beta+1}](1 + r) \tag{14'}$$
$$= \boldsymbol{\pi}(1 + r)$$

Equation 15 is the familiar Leslie-Bernardelli discrete-time version of the Lotka system. [See Keyfitz[31] for a useful account.] Equation 14' is the matrix version of (14). Clearly its solution requires prices to be row-vector eigenfunctions of L and $1 + r$ to be the Malthus-Lotka-Fisher eigenvalue of asymptotic growth. Except for inessential scale coefficient (14') has the following unique solution, which is Fisher's solution adapted to discrete time periods

$$[v_1 \, v_2 \cdots v_\beta 0]\mathbf{L} = [v_1 \, v_2 \cdots v_\beta 0](1 + m)$$

$$v_1 = 1 = \sum_{1}^{\beta+1} (p_1 \cdots p_{j-1})b_j(1 + m)^{-j} = \sum_{1}^{\beta+1} \ell_j b_j(1 + m)^{-j}$$

$$\cdots \cdots \cdots \cdots$$

$$v_x = \sum_{x}^{\beta+1} (\ell_j/\ell_x)b_j(1 + m)^{-j+x-1}, \quad 1 \le x \le \beta + 1 \tag{16}$$

$$\cdots \cdots \cdots \cdots$$

$$v_\beta = b_\beta(1 + m)^{-1}$$

$$v_{\beta+1} = 0$$

The subtle reader will protest: In (15) you made your quantity iterations for $N(x, t)$ dynamic. Why did you not do the same in (14′) for the *dual* price vector $\bar{\pi}_x(t)$, replacing (14′) by

$$[\bar{\pi}_1(t + 1) \cdots \bar{\pi}_\beta(t + 1)] \begin{bmatrix} b_1 & \cdots & b_{\beta-1} & b_\beta \\ p_1 & \cdots & 0 & 0 \\ \cdot & & \cdot & \cdot \\ \cdot & & \cdot & \cdot \\ \cdot & & \cdot & \cdot \\ 0 & \cdots & p_{\beta-1} & 0 \end{bmatrix} \tag{15′}$$

$$= \bar{\pi}_x(t + 1)\bar{L} = \bar{\pi}_x(t)$$

That reader has a point: since the population is for a long time in transient nonsteady states, why should the steady-state price be appropriate during those states?

The best answer I can give, and it seems a somewhat feeble one, is this. There is only one normalized initial $\bar{\pi}_x(0)$, namely that chosen to be proportional to [1, $v_2, \ldots, v_\beta$], such that the iteration

$$\bar{\pi}_x(t + 1)\bar{L} = \bar{\pi}_x(t) \tag{16}$$

$$\bar{\pi}_x(t + 1) = \bar{L}^{-1}\bar{\pi}_x(t) \tag{17}$$

does not for some finite t give solutions with some negative components. [See Wan[32] for discussion of these so-called Furuya-Inada and F. Hahn properties of permanently extendable equilibrium systems.]

If anything, our admiration grows for Fisher's blindfolded gallop over eggshells. His sureness of foot in reaching useful results is astounding. Those of us not geniuses had better not shut our eyes in exploring such treacherous new ground. The following section shows the need to discipline intuition by logical proof.

To apprehend the subtle general (lack of!) logical connection between correct steady-state economic prices and true reproductive value of initial population elements, consider a rockbottom simple case of a biparental population model. Babies of both sexes are created in equal numbers; their mothers are all young women; their fathers are young or less-young men, but of course the latter display less fertility than the former. With the age-specific-mortality of men specified, and also the fertility function specified in terms of the 3 arguments—young women, $F(t)$, young men $M(1,t)$, and less-young men, $M(2,t)$,—we have a determinate nonlinear dynamic system that will approach an asymptotic mode of balanced exponential growth. Here, as in Yellin and Samuelson[33] are its dynamic relations

$$\tfrac{1}{2}B(t+1) = F(t+1) = M(1, t+1) = b[F(t), M(1, t) M(2, t)] \qquad (18.1)$$

$$M(2, t+1) = \ell M(1, t), 0 < \ell \le 1$$

$$b[\lambda x, \lambda x_2, \lambda x_3] \equiv \lambda b[x_1, x_2, x_3], \partial b/\partial x_i = b_i[x_1, x_2, x_3] > 0 \qquad (18.2)$$

$$\sum_1^3 x_j b_j[x_1, x_2, x_3] \equiv b[x_1, x_2, x_3]$$

$$\lim_{t \to \infty} \frac{B(t)}{(1+m)^t} = v[F(0), M(1, 0), M(2, 0)]/[1 + \tfrac{1}{2}\ell] \qquad (18.3)$$

where $1 + m = M^*$ is the unique positive root for M in the steady-state relation

$$b[1, 1, \ell M^{-1}] - M = 0, \quad 0 < 1 + m \gtrless 1 \qquad (18.4)$$

The question is now well-posable. How is the true reproductive-value function, $v[F(0), M(1,0), M(2,0)]$, related—if at all—to the correct steady-state prices applied to the initial elements of the population?

In Fisher's exposition, this question is answered in an affirmative and quantitative manner. But, as we shall now see, in general biparental models it is *not* the case that calculating present discounted values of the charter members of the population, at the true asymptotic Malthusian growth rate or at *any* other interest rate, will give the true reproductive value.

I now calculate these present discounted values in what modern advanced economics has shown to be the only way that they can be defined. First, we must calculate the rents (in terms of babies of course) that each person of each sex will earn during each period of life. Modern advanced economics tells us that when 2 or more inputs cooperate in producing an output [strictly speaking, by a homogeneous-first-degree, differentiable function like $b[x_1, x_2, x_3]$ in (18)], the separate contribution that can be imputed to each is identifiable by the respective partial derivative of the production function (its so-called "marginal productivity"). Hence, at *any* time, which may be the *original* time and need not be a time in which the system has settled into its long-term stable-age distribution steady state, these rents are given by

$$q_F(t) = \frac{\partial b[F(t), M(1, t), M(2, t)]}{\partial F(t)} \qquad (19)$$

$$q_{M_i}(t) = \frac{\partial b[F(t), M(1, t), M(2, t)]}{\partial M(i, t)}, (i = 1, 2)$$

The steady-state rents are therefore

$$q_F^* = b_1[1, 1, \ell(1 + m)^{-1}]$$

$$q_{M_1}^* = b_2[1, 1, \ell(1 + m)^{-1}] \qquad (20)$$

$$q_{M_2}^* = b_3[1, 1, \ell(1 + m)^{-1}]$$

The steady-state present-discounted-values of each age-and-sex group would then be

$$v_F^* = b_1[1, 1, \ell(1 + m)^{-1}](1 + m)^{-1}$$

$$v_{M_2}^* = b_3[1, 1, \ell(1 + m)^{-1}](1 + m)^{-1} \tag{21}$$

$$v_{M_1}^* = b_2[1, 1, \ell(1 + m)^{-1}](1 + m)^{-1} + b_3[1, 1, \ell(1 + m)^{-1}](1 + m)^{-2}$$

In the last row above, we have to take proper notice that a young man will father babies in 2 stages of life, and we must discount more heavily the fruit that will come only later.

Any numerical example will verify that true total reproductive value, such as is well defined in (18.3), is *not* given by a Fisher-like calculation of

$$v_F^* F(0) + v_{M_1}^* M(1, 0) + v_{M_2}^* M(2, 0) \tag{22}$$

Why not? Because the charter member's own fruit is *not* that which they'd earn working with numbers in the other groups that are in the ultimate balanced ratios, but rather what they contribute when working with the *actual initial* unbalanced ratio.

What would be the present discounted values of the original groups calculated at their *actual* future economic rents? Evidently, the following obtains at *any* t:

$$v_F(t) = b_1[F(t), M(1, t), M(2, t)] (1 + m)^{-1}$$

$$v_{M_2}(t) = b_2[F(t), M(1, t), M(2, t)] (1 + m)^{-1} \tag{23}$$

$$v_{M_1}(t) = b_3(F(t), M(1, t), M(2, t)) (1 + m)^{-1} + \ell v_{M_2}(t + 1)(1 + m)^{-1}$$

Again, it is *not* the case that the following equality holds:

$$V(t + 1) = (1 + m)V(t) \tag{24}$$

$$V(t) = v_F(t)F(t) + \sum_{1}^{2} V_{M_j}(t)M(j, t) \tag{24.2}$$

However, and here my quill takes on Fisherine style, it can be seen (or shown) that

$$\lim_{t \to \infty} [B(t)(1 + m)^{-t}/V(t)] = 1 \tag{25}$$

So to speak, penultimately, use of current present discounted values (or even discounted values that employ steady-state prices) will give a "good approximation" to true reproductive value. But I fear this is the case of a candle that gives a light near the bright noon of day when we need it least. If the system is already in its penultimate state *near* the stable-age-distribution configuration, it suffers little from transient oscillations in its relative age distribution and we have no great need to damp them down.

The reader may examine a nonlinear case already known to be soluble and satisfy himself that applying Fisherine present-discount-value formulas in cookbook fashion will not procure the desired soufflé of true reproductive value.

I may sum up these deeper economic investigations of biology thus. Fisher did not prove, nor even conjecture clearly, that present discounted value calculated with steady-state or other prices does (1) provide a scalar that grows from the beginning at the ultimate Malthusian rate, (2) provide a proper measure of an initial age element to the system's ultimate growth level. In general, such a conjecture would be false. Like Tartaglia (who had a formula for the roots of one kind of polynomial, the cubic), and unlike Sturm or Gauss (who knew how to calculate all the roots of any polynomial, and knew when there existed no closed-form formula for doing so), R. A. Fisher glimpsed the rule valid [only!] for the Lotka one-sex linear model and stated that rule in anagramic mystery. Fisher's command over his readers is revealed by their docile acceptance for half a century of his abracadabra. (Perhaps he was his own first conquest.)

Conclusion

It will not escape Robert K. Merton's attention—I insist on the "K." since he and I have a "C." in common—that the present exercise also belongs in his zoo. Pioneering work, Littlewood and Hardy agreed, is usually clumsy. It is hard to be elegant while conquering Everest for the first time. Later, Lilliputian tailors provide a more elegant dress or axiomatization. Often that dress is a shroud, of more interest to the embalmer and taxidermist of science—its philosopher, historian, and sociologist—than to its card-carrying tillers. Still, as they say down at the census, even undertakers' services are part of the Gross National Product, along with those of the midwife.

Acknowledgments

I have benefitted from joint researches with Joel Yellin of Massachusetts Institute of Technology, and from writings and conversation of Nathan Keyfitz of Harvard. Kate Crowley gave much appreciated editorial assistance. Robert K. Merton alerted me to what might otherwise have been taken for granted. After this paper was set up in type, I was able to read the fine biography of R. A. Fisher by his daughter, Joan Fisher Box.[34] It corrects my surmise that Fisher's interest in biology post-dated his undergraduate days. But, with qualifications, I find it confirming of my general conclusions.

References

1. FISHER, R. A. 1958. *Genetical Theory of Natural Selection.* Dover Publications, Inc. New York. Second revised edition of 1930 Oxford University Press first edition.
2. NEYMAN, J. 1935. Statistical Problems in Agricultural Experimentation. *J. R. Stat. Soc.* **2:** 107–154.
3. MERTON, R. K. 1973. *The Sociology of Science.* The University of Chicago Press, Chicago.
4. FISHER, R. A. Ref. 1. In all cases I quote from the 1958 revision; but, actually, as is readily apparent from study of the smaller type in which changed lines were set, no substantial changes from the 1930 text occur in any of the quoted passages.
5. KUCZYNSKI, R. R. 1928. *The Balance of Births and Deaths,* Vol. I. Macmillan. New York.
6. SHARPE, F. R. & A. J. LOTKA. 1911. A Problem in Age Distribution. *Philosophical Magazine* **21:** 435–438.
7. LOTKA. A. J. 1913. A Natural Population Norm. *J. Wash. Acad. Sci.* **3:** 241–248; 289–292.
8. LOTKA. A. J. 1956. *Elements of Mathematical Biology.* Dover Press. New York. See this publication for a list of relevant pre-1930 Lotka writings.
9. LOTKA, A. J. 1925. *Elements of Physical Biology.* Wilkins and Wilkins. Baltimore. This is the original edition from which the amplified 1956 posthumous Dover edition of reference 8 was derived.

10. FISHER, R. A. 1934. Randomisation, and An Old Enigma of Card Play. *Mathematical Gazette* **18**:92–297.
11. VON NEUMANN, J. 1928. Zur Theorie der Gessellschaftsspiele. *Mathematische Annalen* **100:** 295–302.
12. SEAL, H. L. 1967. The Historical Development of the Gauss Linear Model. *Biometrika* **54:** 1–24. The quotations are taken from p. 2.
13. FISHER, R. A. Ref. 1: 23–26.
14. WILSON, E. O. 1975. *Sociobiology*. Harvard University Press. Cambridge, Mass.
15. SAMUELSON, P. A. 1976. Resolving a Historical Confusion in Population Analysis. *Hum. Biol.* **48:** 559–580. Equation 19. Reproduced as Chapter 236 in *The Collected Scientific Papers of Paul A. Samuelson,* Vol. 4. H. Nagatani & K. Crowley, Eds. 1978. The MIT Press. Cambridge, Mass.
16. LOPEZ, A. 1961. *Problems in Stable Population Theory*. Office of Population Research. Princeton, N.J.
17. KEYFITZ, N. 1977. *Applied Mathematical Demography*. John Wiley & Sons. New York.
18. SAMUELSON, P. A. 1977. Generalizing Fisher's "Reproductive Value": Linear Differential and Difference Equations of "Dilute" Biological Systems. *Proc. Nat. Acad. Sci. U.S.A.* **74:**5189–5192.
19. SAMUELSON, P. A. 1977. Generalizing Fisher's "Reproductive Value": Nonlinear, Homogeneous, Biparental Systems. *Proc. Nat. Acad. Sci. U.S.A.* **74:** 5772-5775.
20. SAMUELSON, P. A. 1978. Generalizing Fisher's "Reproductive Value": Overlapping and Nonoverlapping Generations. *Proc. Nat. Acad. Sci. U.S.A.* **75:** 4062–4066.
21. SAMUELSON, P. A. 1978. Generalizing Fisher's "Reproductive Value": "Incipient" and "Penultimate" Reproductive-Value Functions When Environment Limits Growth; Linear Approximants for Nonlinear Mendelian Mating Models. *Proc. Nat. Acad. Sci. U.S.A.*
22. DUBLIN, L. I. & A. J. LOTKA. 1930. *Money Value of a Man*. The Ronald Press Company. New York.
23. WALSH, J. R. 1936. The Capital Concept Applied to Man. *Q. J. of Econ.* 255–285.
24. FOGEL, R. W. & S. L. ENGERMAN. 1974. *Time on the Cross*. Little Brown. New York.
25. SAMUELSON, P. A. 1958. An Exact Consumption-Loan Model of Interest with or without the Social Contrivance of Money. *J. Polit. Econ.* **66:** 467–482. Reproduced as Chapter 21 of *The Collected Scientific Papers of Paul A. Samuelson,* Volume 1. J. E. Stigilitz, Ed. 1966. The MIT Press. Cambridge, Mass.
26. PHELPS, E. S. 1966. *Golden Rules of Economic Growth*. W. W. Norton. New York.
27. SAMUELSON, P. A. 1937. Some Aspects of the Pure Theory of Capital. *Q. J. Econ.* **51:** 469–496. Reproduced as Chapter 17 of *The Collected Scientific Papers of Paul A. Samuelson,* Vol. I. J. E. Stiglitz, Ed. 1966. The MIT Press. Cambridge, Mass.
28. BOULDING, K. 1934. Applications of the Pure Theory of Population Change to the Theory of Capital. *Q. J. of Econ.* **45:** 645–666.
29. LESLIE, P. H. 1945. On the Use of Matrices in Certain Population Mathematics. *Biometrika* **33:** 183–212.
30. VON NEUMANN, J. 1945. A Model of General Economic Equilibrium. *Review of Economic Studies* **13:** 1–9. Translation of 1931 and 1936 German version.
31. KEYFITZ, N. 1968. *Introduction to the Mathematics of Population*. Addison-Wesley Publishing Co. Reading, Mass.
32. WAN, H. 1971. *Economic Growth* (Chapter 11). Harcourt Brace Jovanovich. New York.
33. YELLIN, J. & P. A. SAMUELSON. 1977. Comparison of Linear and Nonlinear Models for Human Population Dynamics. *Theor. Popul. Bio.* **11:** 105–126.
34. BOX, J. F. 1978. *R. A. Fisher: The Life of a Scientist*. John Wiley & Sons. New York.

MERTON ON MULTIPLES, DENIED AND AFFIRMED

George J. Stigler

Charles R. Walgreen Foundation
Department of Economics
University of Chicago
Chicago, Illinois 60637

A science is conducted by a society of scholars who jointly pursue the development of a coherent body of knowledge, including a central theoretical core. This society, using practices such as specialization and exchange which we associate with economic societies, tests received doctrines, extends their applicability, and strives to discover the explanations of phenomena presently inexplicable with received doctrines. Because the pursuit of science is a social enterprise, only knowledge shared by its members is scientific. As in other social enterprises—economic, military, political—it is superficial and misleading to view the progress of the society as the product of a few heroic figures.

No one has done more to develop the implications of this view of science and scientists than Robert K. Merton—indeed no one else has done anywhere near so much to develop the study of science as a social enterprise.[1] (Merton's contributions have been so fundamental as to constitute almost a self-refutation of his thesis of science as a social enterprise!) Among the implications which he has drawn, none is more telling than his thesis that "all scientific discoveries are in principle multiples, including those that on the surface appear to be singletons."[2] This essay examines this thesis, viewed as an explanatory principle for the scientific discoveries in economics.

THE THESIS OF MULTIPLES

Rather than give a formal statement and proof of the thesis, Merton presents ten kinds of phenomena and behavior in science that are implicit or explicit suggestions that the probability is high that any important scientific discovery will be made by more than one person.[3]

These various evidences for multiples include the discoveries that proved to have one or more complete anticipations, the announcements of duplicated researches or of nearly completed researches abandoned because of others' discoveries, the race to publish new results, and the attempts to achieve and protect priority of discovery. An inventory of multiple discoveries made in collaboration with Dr. Elinor Barber is briefly discussed.

We can produce a list of multiples in economics, the first of which was surely in Merton's list, for he quotes its mention by Macaulay:

1. ". . . the doctrine of rent, now universally received by political economists, was propounded, at almost the same moment, by two writers unconnected with each other." They were T. R. Malthus and E. West (1815).
2. The near simultaneous discovery of the marginal utility theory by Jevons (1862), Menger (1871) and Walras (1874).
3. The marginal productivity theory, discovered by Marshall (1879), Edgeworth (1881), Stuart Wood (1888), Wicksteed (1894), as well as Barone (1895), J. B. Clark (1889), and no doubt others.

143

0028–7113/80/0039–0143 $1.75/2 © 1980, NYAS

4. Monopolistic and imperfect competition, discovered by Chamberlin (1934) and J. Robinson (1932).[4]
5. The modern theory of utility, including the Slutsky equation, discovered by Slutsky (1915) and Hicks and Allen (1934).
6. The theory of comparative cost, due to Ricardo (1817) and Torrens (1815).
7. Refutation of the wages-fund theory by W. Thornton (1869) and Francis Longe (1866).
8. The international factor equalization theorem, due to Lerner (1933) and Samuelson (1948).

In addition, many minor multiple discoveries are known.[5] Examples are the demonstration of the existence of a measurable utility function when the utility function is additive in its arguments (Wicksteed, 1888; Fisher, 1892), and the discovery of the kinked oligopoly demand curve by Paul Sweezy (1939) and Hall and Hitch (1939).

THE RATIONALE OF MULTIPLES

In the basic essay on multiple discoveries, Merton does little more than hint at the reason for their existence. Of the caveats that bring this important essay to a conclusion, one states that multiple discoveries need not be chronologically simultaneous: two discoveries can be "simultaneous or nearly so in social and cultural time, depending upon the accumulated state of knowledge" in the several cultures in which they appear.[6]

In a later essay, "Multiple Discoveries as Strategic Research Site," these hints are expanded upon:

> The sheer fact that multiple discoveries are made by scientists working independently of one another testifies to the further crucial fact that, though remote in space, they are responding to much the same social and intellectual forces that impinge upon them all.[7]

In this essay a wholly different scientific function is also assigned to multiple discoveries:

> Often a new idea or a new empirical finding has been achieved or published, only to go unnoticed by others, until it is later uncovered or independently rediscovered and only then incorporated into the science. . . . Multiples—that is, redundant discoveries—have a greater chance of being heard by others in the social system of science and so, then and there, to affect its further development.[8]

The argument that the discoveries are called for by the preceding development of the science is surely the essential basis for expecting multiple discoveries. Previous scientific evolution has thrown up problems on methods or principles which make the succeeding discoveries necessary for continued scientific work. The rent theory of West and Malthus (and Ricardo) was appropriate to an island economy in which a rapidly growing population and rapidly expanding industrial production would put progressively stronger strains on the capacity of the domestic agricultural system. The Slutsky equation presented a fundamental relationship immanent in the theory of utility-maximizing behavior.

But if multiple discoveries are a response to generally felt scientific needs, we must *define* multiple discoveries as those which appear at a given stage in the evolution of a science. If important elements of Keynesian economics were discovered by Kalecki in Poland in the 1930s, we should hardly call this a multiple because the ruling (Marxian) economics of Poland bore little relationship to that of Great Britain.

The determination whether the state of a science in country A is the same as that in country B, either at the same or different times, is not an easy task but it is not an impossible one. If the scientists in A and B are working on the same problems, perhaps citing the same literature, they also share the implicit need for the scientific discoveries which are necessary for further progress.[9] For example, was Turgot's statement of diminishing returns (1767) a multiple with that of Malthus and West in England 48 years later? I would say not, because no use was made of Turgot's theory by the Physiocrats, who were not concerned with the effect of agricultural protection on prices and incomes, as the English economists were. On the other hand, Irving Fisher and Wicksteed were concerned with the same theory of utility in their proposals to measure utility with additive functions.

There is no evidence that Merton has systematically imposed a test of similarity of scientific environment in the determination of multiple discoveries. When we use the test, many multiple discoveries will vanish. Let us reconsider our short list from economics.

1. The theory of rent and diminishing returns *is* a true example: all three economists (for Ricardo should be included) were writing in the same scientific setting.
2. The marginal utility theory is a much more dubious example: the status and direction of economic science in France, Switzerland and Vienna were rather different than in England.
3. The marginal productivity theory becomes a much lower multiple—Wood in America and Barone in Italy, for example, are in somewhat different scientific contexts than the British economists.
4. Monopolistic and imperfect competition is a true multiple in settings, but the two theories differ in fundamental respects.
5. Slutsky writing in Russia in 1915 and Hicks and Allen in Britain in the 1930s are in quite different scientific worlds.
6. Comparative cost theory is a true multiple.
7. The refutation of the wages fund is a true multiple, but marred by the fact that the refutation was idle until a superior theory (marginal productivity) appeared somewhat later.
8. The factor equalization theorem was perhaps a multiple, although the fact that the earlier version (developed at a major center of economics) was not published raises perplexing questions.

In this very brief list, about half of the multiple discoveries appear to survive the essential requirement that they were made in similar scientific settings.

But let us now come to our main point: most of the multiple discoveries were not multiples at all, *but they nevertheless support Merton's basic thesis.* For most of the multiples were discoveries that had been made earlier but had been ignored:

1. James Anderson (1777) had the rent theory (without diminishing returns in the modern sense); Turgot stated the law of diminishing returns in 1767.
2. The marginal utility theory had many early discoverers, of whom Gossen (1854) was most noteworthy.
3. The marginal productivity theory was proposed by Longfield (1832) and von Thünen (1850).
4. The theories of Chamberlin and Joan Robinson (which differ in important respects) are largely anticipated by Marshall and (later) by Sraffa.
5. Slutsky should be looked upon as an anticipator of Hicks and Allen.
6. Comparative cost theory had several incomplete anticipations (Viner, *Studies in the Theory of International Trade,* p. 440).
7. The refutation of the wages fund theory (by denying the fixity of the fund) is not credited to any earlier writers, although I am utterly confident that the

main point (that the wages fund was not a fixed quantity) was common in the radical literature.

8. The factor equalization theorem lacked an earlier anticipation.

In fact, we may reasonably expect most of the multiples that survive the condition of similar scientific settings to be anticipated by earlier, less successful discoveries of the same ideas.

The unsuccessful earlier discoveries are the very evidence for the "inevitability" of scientific progress that the multiple discoveries was supposed to present. If an early, valid statement of a theory falls on deaf ears, and a later restatement is accepted by the science, this is surely proof that the science accepts ideas only when they fit into the then-current state of the science. Gossen, writing in the high tide of German Historical economics, was simply inappropriate to his scientific environment. Longfield in Ireland, and von Thünen in Germany, were presenting a marginal productivity theory for which neither German nor British economic science was ready. And similarly for Slutsky, Cournot and other unsuccessful discoverers.

On this view, Merton's secondary task for multiples of persuading a science to adopt the idea is incompatible with this basic theory. If the science is ready for an idea, it will rarely need multiple discoverers to persuade it to adopt the idea. Elaboration, repetition, and controversy will be the main vehicles of persuasion. The multiples arise because they are demanded by the evolving science.

CONCLUSION

On the present interpretation, multiple discoveries are indeed evidence that the advancing frontier of a science requires new analytical (or empirical) armament and the demand is being met by able scholars. But there is only one reason why full multiples—completed and tested—should occur and that is incomplete knowledge of who is working on a problem and what his achievement will be. The better the information network of a science, the fewer will be true multiples that are separated by a significant period of time.

The unsuccessful early discoveries are equally valid evidence for the social character of science. Indeed it is probable that they are more frequent than multiples. If so, we have completed the full circle: Merton's fundamental thesis is reaffirmed, but multiple discoveries shrink to a minor support for the thesis.

REFERENCES

1. MERTON, R. K. 1973. *The Sociology of Science*. University of Chicago Press, Chicago. Much of his work is assembled in this volume.
2. MERTON, R. K., Ref. 1:356.
3. One would expect the thesis to hold even more fully for unimportant scientific discoveries. I conjecture that the lesser interest and the greater difficulty of enumerating such discoveries has kept them out of the discussion.
4. One must apologize to Chamberlin, who spent much of his life explaining the difference between the two concepts, for combining them here.
5. Mark Blaug has called to my attention another significant (and debatable) multiple; see J.V. Pinto, "Launhardt and Location Theory: Rediscovery of a Neglected Book," *Journal of Regional Science,* 17 (1977): 17–30.
6. MERTON, R. K., Ref. 1:369.
7. MERTON, R. K., Ref. 1:375.
8. MERTON, R. K., Ref. 1:380.
9. This implies that two scientists in the same country at the same time would *not* make multiple discoveries if they were working in well-separated specialties.

STIGLER'S LAW OF EPONYMY*

Stephen M. Stigler

*Department of Statistics
University of Chicago
Chicago, Illinois 60637*

No reader of Robert K. Merton's work on the reward system of science could fail to be struck by his insightful and engaging discussions of the role of eponymy in the social structure of science. The uninitiated should read (and reread) his 1957 address, "Priorities in Scientific Discovery,"[1] but for present purposes I must at least repeat his definition of eponymy, as "the practice of affixing the name of the scientist to all or part of what he has found, as with the Copernican system, Hooke's law, Planck's constant, or Halley's comet."[2] Merton went on to discuss three levels of a hierarchic order of eponymous practice: at the top there are a few men for whom an entire epoch is named, then comes a larger number of scientists designated as "father" of a particular science, and, finally, "thousands of eponymous laws, theories, theorems, hypotheses, instruments, constants, and distributions."[3] The present paper is an attempt by an Outsider to the sociology of science to shed some light on the workings of the eponymic reward system at this third level, and a report on a small statistical investigation into eponymous practices of my own field, statistics.

I have chosen as a title for this paper, and for the thesis I wish to present and discuss, "Stigler's law of eponymy." At first glance this may appear to be a flagrant violation of the "Institutional Norm of Humility,"[4] and since statisticians are even more aware of the importance of norms than are members of other disciplines, I hasten to add a humble disclaimer. If there is an idea in this paper that is not at least implicit in Merton's *The Sociology of Science,* it is either a happy accident or a likely error. Rather I have, in the Mertonian tradition of the self-confirming hypothesis, attempted to frame the self-proving theorem. For "Stigler's Law of Eponymy" in its simplest form is this: "No scientific discovery is named after its original discoverer."

Examples affirming this principle must be known to every scientist with even a passing interest in the history of his subject; in fact, I suspect that most historians of science, both amateur and professional, have had their interest fueled early in their studies by the discovery (usually accompanied by an undisguised chortle) that some famous named result was known (and better understood) by a worker a generation before the result's namesake. A detailed study of any scientific area will show, I would argue, that this phenomenon persists with a generality rivaling that of any other "law" in the social sciences, indeed even that of Merton's famous hypothesis that "all scientific discoveries are in principle multiples."[5]

Merton's hypothesis is related to, yet distinct from Stigler's Law (henceforth humbly referred to as simply the Law). It might appear that the Law is in fact stronger than the hypothesis, that the Law states that a discovery is always named after the wrong one of its multiple discoverers. But this is not a consequence of the Law; a discovery may in fact be named after someone who could not be reasonably counted as even one of its discoverers, much less the original one. Thus a scrupulous examination of the works of economist Robert Giffen has failed to reveal even a

*This work was supported by grants from the National Science Foundation, Nos. SOC 78-01668 and BNS 76-22943 A02.

0028-7113/80/0039-0147 $1.75/2 © 1980, NYAS

semblance of a statement (much less a proof) of what has come to be known commonly as "Giffen's paradox," although an earlier statement (published before Giffen's birth by Simon Gray) has been noted.[6] And St. Matthew did not discover the Matthew effect!

Evidence in favor of the Law is readily available in any field whose history has been subjected to serious scrutiny. Thus in my own field of mathematical statistics it can be found that Laplace employed Fourier transforms in print before Fourier published on the topic, that Lagrange presented Laplace transforms before Laplace began his scientific career, that Poisson published the Cauchy distribution in 1824, 29 years before Cauchy touched on it in an incidental manner, and that Bienaymé stated and proved the Chebychev inequality a decade before and in greater generality than Chebychev's first work on the topic. (Incidentally, in each of these cases there is evidence, sometimes even citation, to show that the earlier work was known to the later worker before he embarked on his investigation. These were *not* instances of multiple discovery.) Examples of this type are not in short supply, and they are not all cases where the discovery preceded its namesake: "Mayer's method" of combining inconsistent linear equations really first appeared in work of Laplace published a quarter century after Mayer died, and recent scholarship[7] has shown that one of the most famous of mathematical relations, the Pythagorean Theorem, was known before Pythagoras, was first proved after Pythagoras, and in fact Pythagoras himself may have been unaware of the geometrical significance of the theorem! But while such examples and other anecdotal evidence could be multiplied with ease,* a true defense of the Law would require a much more arduous examination of the rather ill-defined population of eponyms than I am prepared to undertake at present.[8] Instead, I shall accept the Law as true, and I shall concentrate on adducing reasons for its universality and implications for the reward system of science.[9]

One explanation for the Law has been given by a historian of science in these words: "Every scientific discovery is named after the last individual too ungenerous to give due credit to his predecessors." (That I do not identify the source of this quotation is due to a lack of information, not a lack of generosity.) This analysis of eponymic inaccuracy is witty, but surely false. Strictly interpreted it would imply that discoveries never receive lasting names (since ungenerosity shows no sign of being extinct), and such a claim would be easily refuted, witness the aforementioned Pythagorean Theorem, and Pascal's arithmetical triangle (which was actually published earlier by Pascal's teacher Hèrigone,[10] and was known in China before that). Even if we loosely interpret the statement, as blaming inaccurate names upon inadequate citation and a lack of corrective historical scholarship,[11] it is wrong: frequently posterity has pinned a label on a discovery despite the honored individual's citation of a worthy predecessor, or in the face of abundant historical evidence suggesting another candidate, as we shall see later.[12]

It is also not true that eponyms are bestowed capriciously. They are, as I have claimed, inaccurate as a means of identifying a discovery's originator, but it is rare that an eponym is awarded to an individual who has not done some work at least tangentially connected with the discovery, and rarer still that he has not made important contributions to his science generally. If the Law is not due to ignorance or caprice, and if other explanations such as stupidity or deceit are dismissed out of hand, to what *is* it to be ascribed? I wish to argue that the inability of eponymical practice to

---

*In fact, the Law is at once exemplified and self-exemplified in this statement from G. J. Stigler, *The Theory of Price,* 3rd edition (New York: Macmillan, 1966), page 77: "Here the P represents Hermann Paasche, who, like Laspeyres, was not the first to propose the index named after him. If we should ever encounter a case where a theory is named for the correct man, it will be noted."

meet the assumed purpose for the practice (to commemorate a discovery's original discoverer) is in fact a necessary consequence of the real role the practice plays, which Merton has taught us is as a key element of the reward system of science.

I begin with two observations. First, names are not given to scientific discoveries by historians of science or even by individual scientists, but by the community of practicing scientists (most of whom have no special historical expertise). Second, names are rarely given, and never generally accepted unless the namer (or accepter of the name) is remote in time or place (or both) from the scientist being honored. I shall present some evidence relevant to these claims (particularly the crucial second observation), but let us first pause to consider why they are true, and how they are related to the reward function of eponyms.

The most prestigious eponyms stand at the pinnacle of the scientific reward system—a scientist's name is enshrined in the literature as a mark of the enduring significance of his work, promising to remain there long after his work has ceased to be directly cited by the profession;[13] a kind of intellectual immortality is achieved. If these statements are to be true (and they must be widely seen as true or the eponym would cease to function as an important scientific reward), then the award of an eponym must not only be made on the basis of the scientific merit of originality, but more importantly it must be perceived by the community of scientists as based on merit and not upon personal friendship, national affiliation, or the political pressures of scientific schools. Historians of science may provide lists of nominations for eponymical recognition, but if an eponym is to be viewed as meritorious, then the community will look to specialists in the area of the discovery for guidance, not to historians who are usually specialists in no area. But more must be true—the scientists whose works are consulted for approval of the eponym must be seen as impartial, as only swayed by scientific judgment. An award of an eponym may be attempted by close friends, students, or political associates, but it will not be successful. It is the acceptance by the community at a distance, and thus the promise of immortality through acceptance by future generations of scientists, that gives the award its extraordinary prestige.

Some scientists, when first confronted with the Law, pause only briefly before reciting a string of supposed counterexamples. Many of these examples can be shown on further examination to be confirmations of the Law (although lengthy research may be necessary), but others fall in one of two general categories that would require separate handling in a more definitive investigation of this topic. Eponyms can be found in a wide variety of sizes (from the "F-statistic" to the "Fisher-Neyman-Halmos-Savage factorization theorem"), and in many flavors (from the redundantly reverential "Gaussian linear model" to the accusatory "so-called Cauchy distribution"[14]), and it would be impossible for any simply stated theory to accommodate all such current usage. One large category of examples that appear eponymous at first glance (and because of the close proximity in time and space to the individual named also appear to be exceptions to the Law) are in reality a short form of citation: as common knowledge of the (often important) cited article fades with the passage of time, so does the use of the "eponym," to be replaced by a more specific journal citation, where needed. The Law is not intended to apply to usages that do not survive the academic generation in which the discovery is made. I do not mean to exclude the possibility that an eponym may be contemporary to the discovery it names, although such cases are rare and according to my second observation the namer would have to be remote from the honored scientist in place or discipline. I do, however, insist that an eponym demonstrate its widespread acceptance as a name, and the test of time is the simplest way to show this.

Another class of eponyms that would deserve separate treatment in a deeper and more thorough investigation of this subject is that of multiple awards, such as the aforementioned "Fisher-Neyman-Halmos-Savage factorization theorem." This example, which includes mention of two antagonists (at least three if the word is used in a philosophical sense) and salutes work done on two continents over a quarter-century span, is typical of many that achieve the requisite impartiality[15] by a very different route than does the simple eponym. This shotgun approach, which represents a statement that the development of the idea was the product of a community of scientists rather than a single individual, is more likely to hit a scientist who could be classified as the original discoverer (although there would be little agreement as to which of the many this was), and, in any case, the variation over time of the names included in the list renders the study of these cases extremely difficult.

The necessity of the appearance of impartiality to the award, and the apparent agreement that this appearance is best achieved through the distancing of the namer from the honored scientist, accounts for the general reluctance of scientists to propose their colleagues for eponymic recognition and the general resistance of the profession to such attempts. One famous example of this concerned the planet Uranus, discovered by William Herschel in England in 1781. Herschel attempted to name the planet "Georgium Sidus" after his patron King George III, but while it was briefly known as "the Georgian" in England, continental astronomers rejected the name as too narrowly nationalistic (despite its adherence to the Law). Ironically, Lalande in Paris suggested as a solution to the dilemma that the planet be called Herschel. This name enjoyed some currency on the continent (although less so in England), but either because the naming of a planet after a mere mortal was considered unacceptable, or because of the inexorable workings of the Law, "Herschel" eventually yielded to Bode's suggestion of "Uranus."[16]

There is another interesting phenomenon that can be explained by the necessity of the appearance of impartiality. I am aware of some notable instances of challenges to an eponymic award where, curiously, the challenge has been made by a student or countryman of the honored scientist in behalf of a scientist from some other country. For example, the claim that Bienaymé, in 1853, published what is known as the Chebychev Inequality (after an 1867 paper of Chebychev), and that Chebychev was well aware of Bienaymé's work, was most convincingly advanced by Chebychev's illustrious student Markov, and has recently been extensively discussed in a book, one of whose authors was born in the Ukraine.[17] I lack objective evidence as to how general this phenomenon is, but if it is widespread, as I suspect it is, it signals that the resistance to eponymic recognition of close associates may in fact be a norm of scientific behavior, one which serves the role of protecting the practice from degenerating to a regional or factional basis, with the consequent fall in the reward's incentive power.

An extreme case of inaccurate eponymous assignment, and important support for both the Law and the case I present for it, was pointed out to me by Robert K. Merton in a letter. I refer to the common practice of naming scientific units (such as the watt, ohm, and volt) after individuals other than their originators. This practice is particularly significant as it has been fully institutionalized in many cases, through bureaus of standards or nomenclature commissions which guarantee the accuracy of the Law by design. While these commissions seek an appropriate matching of individuals and units, they aim for general commemoration of excellence rather than the labelling of units by their inventors' names. The attaching of old names to new units guarantees the validity of the Law in this case, just as the international character of these bodies is intended to provide the appearance of impartiality required for the acceptance of the name.

If my claims are accepted, if eponyms are only awarded after long time lags or at great distances, and then only by active (and frequently not historically well informed) scientists with more interest in recognizing general merit than an isolated achievement (even a conspicuous one), then the Law can be seen to follow. For it should not then come as a surprise that most eponyms are inaccurately assigned, and it is even possible (as I have boldly claimed) that all widely accepted eponyms are, strictly speaking, wrong. The very inaccuracy of the assignment stands as additional testimony to its remoteness and impartiality, and helps to guarantee its prestige and survival!

In the remainder of this paper I wish to present evidence, through a study of the pattern of acceptance of one eponym in my own field, of the rate and manner in which such names are adopted. In keeping with the Mertonian tradition, the study will be quantitative.

The discovery I have chosen to consider is the probability distribution with density

$$f(x) = \frac{1}{\sqrt{2\pi}} e^{-x^2/2}.$$

This distribution is today most commonly called the "normal distribution" or the "Gaussian distribution," after the great mathematician Carl Friedrich Gauss, who associated it with the method of least squares in his first publication on that topic, in 1809. Of course, the Law tells us that because $f(x)$ is now called the Gaussian distribution and Gauss did in fact study it, he must have been preceded. Indeed he was, for Gauss himself cites Laplace in connection with $f(x)$ in his 1809 book, and indeed Laplace did touch on the distribution as early as 1774.[18] But since a very few modern writers call it the Laplace or even the Laplace-Gauss distribution, we must look further back for its origin. Such a search would be rewarded, as current historical scholarship marks the distribution's origin as being a 1733 publication by Abraham De Moivre.[19] Interestingly enough, De Moivre's work was known by Laplace and Gauss, and his claim as the originator of the distribution is substantiated by the fact that *no* modern writer, as far as I am aware, calls it the "De Moivre distribution."

I selected the distribution $f(x)$ for study for two principal reasons. First, it has occupied a central position in mathematical statistics since at least 1810, and thus has been available for eponymical award for a long time. Second, there are several major candidates for the award, from different countries, and the relationship between nationality and acceptance of the name may be investigated. It is also true that the distribution has (and has always had) popular names such as "error curve" or "normal distribution" which may be used as alternatives to eponyms, and thus the use of an eponym may be regarded as a matter of choice, not necessity.[20]

An indication of the current state of eponymic designation of $f(x)$ can be found by consulting several recently compiled permuted title indexes of statistical journals and research papers. The most massive such index is *Index to Statistics and Probability: Permuted Titles,* published in 1975, edited by Ian C. Ross and John W. Tukey.[21] This index covers the literature through 1966, with the preponderance of titles dating from 1945. I find the index gives 1099 titles referring to $f(x)$ as "normal," "Gaussian," or "Laplace-Gauss," or some variation on these names. Of these, 18% (199) referred to Gauss, and 1% (7) referred to Laplace-Gauss, giving an eponymic total of 19% (206). The remainder of the references were to "normal." Another recent index, *An Author and Permuted Title Index to Selected Statistical Journals,* published in 1970, edited by B. L. Joiner, *et al.*[22] gives an eponymic total of only 11% (42 of 330), but that index is based upon only six Anglo-American journals (mostly from the 1960s), and misses what we shall see is a heavier continental usage of eponyms. A better indication of the worldwide practice can be found from the *Current Index to Statistics,* published annually since 1975.[23] The 1975 volume gives an eponymic total of 28% (44 of 159); the 1976 volume gives a total of 30% (56 of 185). All of these references were to Gauss, none to Laplace. Based upon these data, it seems fair to say that about 20% to 30% of all references to $f(x)$ in the titles of current research papers in theoretical or applied statistics are eponyms, and nearly all of these refer to Gauss. The question then is, how was this level achieved? When, where, and at what rate was $f(x)$ awarded to this Titan of Science?

A permuted title index is not available for earlier literature; even if one were it is probable that changes in the character of the scientific literature generally (e.g., from monograph to research paper), and statistics in particular, would render it useless. I have therefore turned to another, more stable source of information, the textbook. We may expect textbook usage of eponyms to be at once more conservative and more liberal than the literature of active scientific research. Since a textbook, in reaching for a large market, may be expected to reflect the views and practices of only large segments of the scientific community, we may expect that it will be more resistant to new names than the active literature is, waiting until the verdict of the community is in. On the other hand, once a name has been accepted by a sizable fraction of the community, we may expect textbooks to be more generous than the community, by listing the name as one of several alternatives (whereas authors of research papers ordinarily use only a single name).

I selected for study a total of 80 textbooks, covering the period 1816 to 1976. All were what I would classify as statistical texts, although the emphasis varied from least squares for geodesists to correlation for economists and sociologists, and the levels varied from elementary to advanced. All made conspicuous mention of the distribution $f(x)$. The selection was not random. I canvassed my own library, and that at the Center for Advanced Study in the Behavioral Sciences, and I included a number of older texts in the Stanford University library which had not yet been placed in inaccessible "auxiliary storage." If the selection is biased, I believe the bias would be toward the inclusion of books by well-known authors, and toward frequently used, often reprinted texts (although in no case was more than one edition of a work included). Where conscious selection was most heavily exercised (recent English language texts), an attempt was made to choose general texts and avoid overrepresentation of any single orientation.

Each book was classified by country and year of publication, and according to how the author described the distribution $f(x)$. (Several authors used more than one description.) The data are presented in the Appendix, and summarized in TABLE 1. A brief explanation of the grouping of TABLE 1 is in order: All books were classified Eponymic (the author gave at least one eponymic description) or Other (he gave *no* such description). This classification differs slightly from that of the APPENDIX, where Noneponymic means the author gave at least one noneponymic description (but may have included eponymic descriptions as well). The countries of origin were grouped as Germanic (Germany and Austria), French (France and Belgium), Other Continental (Italy, Holland, Romania, and the Scandinavian countries), and Anglo-American (England and the U.S.A.). Most of the eponyms encountered (and all prior to 1920) referred to Gauss.

Perhaps the most striking features of TABLE 1 are the slow rate of acceptance of an eponymic description of $f(x)$, and the difference between the acceptances in the Anglo-American and Continental literatures. The first eponym encountered in the sample was in reference to Gauss in a book by F. R. Helmert published in Germany in 1872, 61 years after Gauss's relevant publication and 17 years after Gauss's death. Perhaps partly because of Helmert's identification with Gauss's homeland, the name was slow in gaining currency. An American text (by T. W. Wright) used it in 1884, but it is more likely that it was J. Bertrand's use of "loi de Gauss" in 1889 that signaled the name was acceptable to the larger community of scientists. The use of this eponym seems to have spread steadily after that, achieving some currency in Italy after the First World War, and being mentioned in all five of the post-Second World War Continental texts examined. The apparent recession in its use in France between the wars may be a confirmation that even the most impartial of scientific awards is not immune to political events.

The rate of eponymic acceptance in the Anglo-American literature, on the other hand, has not undergone any marked change since 1884. I suggest that this lower rate in England and America may be an instance of a generally lower use of eponyms in these countries than on the Continent, but the present data set does not permit the investigation of this hypothesis.

Another interesting aspect of these data is the pattern of references to Laplace, the only candidate alternative to Gauss receiving mention. The earliest mention of a Gauss-Laplace distribution in the sample was in an Italian work in 1920; of the four subsequent mentions of Laplace that were noted, three appeared in France. It would seem that after a century (Laplace died in 1827) French authors felt sufficiently

TABLE 1

EIGHTY BOOKS PUBLISHED 1816–1976, CROSS-CLASSIFIED BY YEAR AND COUNTRY OF PUBLICATION, AND ACCORDING TO WHETHER THE BOOK EMPLOYED AN EPONYMIC USAGE FOR $f(x)$ ("EPONYMIC") OR IT EMPLOYED NO SUCH USAGE ("OTHER")

a. Books published 1816–1884	Eponymic	Other	Total	b. Books published 1888–1917	Eponymic	Other	Total
Germanic	1	2	3	Germanic	3	2	5
French	0	7	7	French	4	0	4
Other				Other			
Continental	0	2	2	Continental	0	4	4
Total				Total			
Continental	1	11	12	Continental	7	6	13
Anglo-American	1	4	5	Anglo-American	1	8	9
TOTAL	2	15	17	TOTAL	8	14	22

c. Books published 1919–1939	Eponymic	Other	Total	d. Books published 1947–1976	Eponymic	Other	Total
Germanic	0	1	1	Germanic	2	0	2
French	1	3	4	French	2	0	2
Other				Other			
Continental	2	2	4	Continental	1	0	1
Total				Total			
Continental	3	6	9	Continental	5	0	5
Anglo-American	2	11	13	Anglo-American	4	10	14
TOTAL	5	17	22	TOTAL	9	10	19

distant from their countryman to advance his name for the award. I personally feel that Laplace's historical link to $f(x)$ is stronger (as well as earlier) than Gauss's, but as yet this eponym has not gained much currency. This may be due to eponymic inertia (it is very difficult to change an established name), or to an eponymic version of the Matthew Effect: the award of the prestigious $f(x)$ to Gauss may simply be a signal of the scientific community's verdict that Gauss was the greater mathematician. In any case, Laplace has in recent years been awarded, as a consolation prize perhaps, eponymic recognition through the spreading acceptance of "Laplace distribution" to mean the less important distribution

$$g(x) = \frac{1}{2} e^{-|x|}.$$

The data set I have presented is quite limited in scope, concerned as it is with but a single discovery in a single field. It does support the thesis that eponyms are only awarded by the scientific community at a considerable distance from the recipient of the award, thus lending plausibility to the case presented for the Law. It is likely that studies of other discoveries in other fields would show considerable variation in the time lag between the discovery and the award, according to the importance of the discovery, the nation involved, and the institutional organization of the field involved.[24] We may expect that in years to come, Robert K. Merton, and his colleagues and students, will provide us with answers to these and other questions regarding eponymy, completing what, but for the Law, would be called the Merton Theory of the reward system of science.

ACKNOWLEDGMENTS

I am grateful to many colleagues for their comments and criticisms in the course of the preparation of this paper. I particularly wish to thank Stephen Cole, William H. Kruskal, and George J. Stigler for their efforts (some in vain) to correct me where they believed I have gone astray.

REFERENCES

1. MERTON, R. K. 1973. *The Sociology of Science: Theoretical and Empirical Investigations.* Edited and with an introduction by Norman W. Storer. U. of Chicago Press, Chicago, Ill. Chapter 14, Priorities in Scientific Discovery, first published 1957.
2. MERTON, R. K. Ref. 1: 298.
3. MERTON, R. K. Ref. 1: 299.
4. MERTON, R. K. Ref. 1: 303.
5. MERTON, R. K. Ref. 1: 356.
6. STIGLER, G. J. 1947. Notes on the History of the Giffen Paradox. *J. Polit. Econ.* **55.** Reprinted as Chapter 14 of *Essays in the History of Economics*, George J. Stigler. 1965. U. of Chicago Press, Chicago, Ill.
7. TIAN-SE, A. 1978. Chinese Interest in Right-Angled Triangles. *Historia Mathematica,* **5:** 253–266.
8. An important source for future eponymic research would be *Eponyms dictionaries index. A reference guide to persons, both real and imaginary, and the terms derived from their names,* by James A. Ruffner (Detroit: Gale Research Press, 1977), but even this large volume is not without its errors or omissions. See, for example, the short review in *Language in Society,* **7** (1978):149.
9. Some of my colleagues, commenting on a draft of this paper, have intimated that the norm of humility is not the only reason for avoiding an autoeponymous designation for the Law. They have ventured the suggestion that the Law is not correct, and even helpfully supplied collections of what are purported to be counterexamples. Now it may in fact be that the literal absolute truth of the Law cannot be defended without occasionally descending to the argumentative depths of adumbrationism (for a discussion of which, see page 20ff. of R. K. Merton's *On Theoretical Sociology,* New York: The Free Press, 1967), or appealing to as yet unnoticed (and earlier) multiple discoveries, but this would not be crucial to the main line of the argument. While exceptions to the Law will only be granted after a struggle and on a case by case basis, all that is really necessary is that the reader grant the frequent truth of the Law, and agree to the unreliability of eponyms as guideposts to original discovery. For this, the examples presented (and appeal to the reader's own experience) should be sufficient proof.
10. DAVID, F. N. 1962. *Games, Gods and Gambling:* 81–82. Hafner Pub. Co., New York, N.Y.
11. What Merton has called the palimpsestic syndrome; see his *On the Shoulders of Giants:* 218–219, 1965. Free Press, New York, N.Y.

12. For another, similar adumbration of the Law, see *The Collected Scientific Papers of Paul A. Samuelson* 1966. J. E. Stiglitz, Ed. Vol. 2, 1503. M.I.T. Press, Cambridge, Mass.

13. COLE, J. R. & S. COLE. 1973. *Social Stratification in Science:* 31. U. of Chicago Press, Chicago, Ill.

14. HEYDE, C. C. & E. SENETA. 1977. *I. J. Bienaymé: Statistical Theory Anticipated:* 90. Springer-Verlag. New York, N.Y.

15. In more standard sociological terminology, what I call impartiality, is called the norm of universalism. See Merton, Ref. 1: 270.

16. NEWCOMB, S. 1878. *Popular Astronomy:* 355. Harper & Bros., New York, N.Y.

17. HEYDE, C. C. & E. SENETA. 1977. Ref. 14: 121–124.

18. Laplace's first encounter with $f(x)$ was as an approximation to the posterior distribution of a probability, a very different use than that Gauss made of $f(x)$. If only the linking of $f(x)$ to least squares is considered, then the earliest publication yet found was by the American Robert Adrain in an obscure magazine dated 1808, a year before Gauss's book appeared.

19. WALKER, H. M. 1929. *Studies in the History of Statistical Method:* 13–19. Reprinted 1975 by Arno Press. New York, N.Y.

20. I am currently engaged in joint research with William H. Kruskal into the history of the use of "normal" in this and other connections. See W. KRUSKAL, 1978. Formulas, Numbers, Words: Statistics in Prose, *In: The American Scholar,* **47**: 223–229, for a brief statement of some of our findings.

21. ROSS, L. C. & J. W. TUKEY. 1975. *Index to Statistics and Probability: Permuted Titles.* Two Volumes. R & D Press. Los Altos, Calif.

22. JOINER, B. L., N. F. LAUBSCHER, E. S. BROWN & B. LEVY. 1970. An Author and Permuted Title Index to Selected Statistical Journals. National Bureau of Standards Special Publication 321. U.S. Department of Commerce, Washington, D.C.

23. *The Current Index to Statistics.* 1975–76. Published annually by the American Statistical Association and the Institute of Mathematical Statistics.

24. Some additional evidence relevant to my case may be found in D. Beaver's"Reflections on the Natural History of Eponymy and Scientific Law," in *Social Studies of Science* (Vol. 6, 1976, pp. 89–98). Beaver shows that as of 1961, eponyms honoring twentieth-century discoveries in physics were far less numerous than those naming earlier scientists, when the different sizes of the scientific communities are allowed for: in Beaver's cross-sectional analysis, the population of eponyms did not seem to be growing exponentially. I take this to be most plausibly explained by a long (say, 30–60 years) average delay in the award of eponyms, such as found in the present longitudinal study, and thus at least consistent with the case presented here. Beaver's different conclusion (that either fundamental discoveries are becoming rarer, or eponymic practice is undergoing a marked change) does not seem to me to be warranted by the data.

APPENDIX

DATA ON 80 BOOKS' EPONYMIC PRACTICES

		Names Used		
Year	Country	Gauss	Laplace	Noneponymic
1816	France			*
1837	France			*
1838	England			*
1843	France			*
1846	Belgium			*
1852	Belgium			*
1860	Germany			*
1867	Italy			*
1869	Belgium			*
1872	Germany	*		*
1874	England			*

APPENDIX (*continued*)

Year	Country	Gauss	Laplace	Noneponymic
1877	Germany			*
1877	U.S.A.			*
1878	France			*
1879	Italy			*
1879	England			*
1884	U.S.A.	*		*
1888	England			*
1889	England			*
1889	France	*		
1892	U.S.A.			*
1892	U.S.A.			*
1896	U.S.A.			*
1896	France	*		*
1897	Germany	*		
1901	England			*
1903	Holland			*
1903	Denmark			*
1906	England			*
1906	Italy			*
1906	Austria			*
1906	Germany	*		
1908	Germany	*		
1908	France	*		
1909	Germany			*
1909	France	*		*
1911	England			*
1912	England	*		*
1917	Denmark			*
1919	England			*
1920	Italy	*	*	*
1921	U.S.A.			*
1921	Italy	*		*
1921	England			*
1921	France			*
1921	Austria			*
1923	U.S.A.			*
1924	France		*	
1924	England			*
1925	U.S.A.			*
1925	England			*
1928	France			*
1928	U.S.A.	*		*
1930	France			*
1931	Sweden			*
1931	Italy			*
1931	U.S.A.			*
1937	U.S.A.	*		*
1937	U.S.A.			*
1939	U.S.A.			*
1939	England			*
1947	U.S.A.			*
1948	France	*	*	*
1950	U.S.A.			*

APPENDIX (*continued*)

Year	Country	Names Used		
		Gauss	Laplace	Noneponymic
1950	U.S.A.	*		*
1952	U.S.A.			*
1956	Austria	*		*
1957	Germany	*		*
1957	France	*	*	*
1960	U.S.A.	*		*
1962	England			*
1963	Romania	*	*	*
1965	U.S.A.			*
1967	U.S.A.			*
1968	U.S.A.			*
1968	U.S.A.			*
1969	U.S.A.	*		*
1970	U.S.A.			*
1970	U.S.A.	*		*
1976	U.S.A.			*

André Cournand was born in Paris (1895) and has served on the faculty of the College of Physicians and Surgeons, Columbia University since 1935, where he is now Professor Emeritus in Medicine and Special Lecturer. He received the B.A. (1913) and the M.D. (1930) from the University of Paris. He has been awarded the Nobel Prize in Medicine and Physiology (with D. W. Richards and W. Forssman, 1956), the Lasker Award (1950), the Gold Medal of the Royal Academy of Medicine, Brussels (1958) and the Academy Medal, New York Academy of Medicine (1966). He is a member of the National Academy of Sciences. His recent publications include *Shaping the Future: Gaston Berger and the Concept of Prospective* (edited with Maurice Lévy, 1973) and "The Code of the Scientist and its Relationship to Ethics" in *Science* (1977).

Ralf Dahrendorf has been Director of the London School of Economics since 1974. Born in Hamburg (1929), he received the Dr. Phil. in philosophy and classics from the University of Hamburg, and the Ph. D. in sociology from the London School of Economics. He has been professor of sociology at the Universities of Hamburg, Tübingen and Constance. He is the author of *Class and Class Conflict* (1957), *Society and Democracy in Germany* (1965) and *The New Liberty* (1975). He is a Foreign Associate of the National Academy of Sciences, and has been a Fellow at the Center for Advanced Study in the Behavioral Sciences.

S. N. Eisenstadt was born in Warsaw (1923). He received the M.A. and the Ph. D. from the Hebrew University of Jerusalem, where he is now Professor of Sociology. He has been a visiting professor at Harvard, M.I.T., Michigan, Oslo, Zurich and Chicago. He is a member of the Israeli Academy of Sciences and Humanities, a Foreign Honorary Member of the American Academy of Arts and Sciences and an Honorary Fellow of the London School of Economics. His books include *Political Systems of Empires* (1963) and *Revolution and the Transformation of Societies* (1978).

Yehuda Elkana was born in Yugoslavia (1934), received the M.Sc. in physics from the Hebrew University of Jerusalem (1966) and the Ph.D. in history of ideas from Brandeis (1968). He has been Chairman of the Department of History and Philosophy of Science at the Hebrew University; he was a Fellow at the Center for Advanced Study in the Behavioral Sciences (1973–4), and a Visiting Fellow of All Souls College, Oxford (1977–8). His publications include *The Discovery of the Conservation of Energy* (1974) and *Toward a Metric of Science* (edited with Joshua Lederberg, Robert Merton, Arnold Thackray and Harriet Zuckerman, 1978). He was appointed director of the Van Leer Jerusalem Foundation in 1969.

Yaron Ezrahi, born in Tel Aviv (1940), is a senior lecturer in the Department of Political Science at the Hebrew University of Jerusalem. He received the B.A. and the M.A. from the Hebrew University, and the A.M. and the Ph. D. from Harvard University. He has been a Fellow at the Center for Advanced Study in the Behavioral Sciences, and is the author of "Political Contexts of Science Indicators" in *Toward a Metric of Science* (edited by Y. Elkana, J. Lederberg, R. K. Merton, A. Thackray and H. Zuckerman, 1978). Since 1974, he has been adviser to the Israeli Academy of Sciences and Humanities.

EUGENE GARFIELD is founder, President and Chairman of the Board of Directors of the Institute for Scientific Information®. He was born in New York City (1925) and received the B.S. in chemistry and the M.S. in Library Science from Columbia University, and the Ph. D. in structural linguistics from the University of Pennsylvania. He is a member of the Rockefeller University Council and of the Board of Directors of *Annual Reviews*. His writings include *Essays of an Information Scientist* (1977) and *Citation Indexing: Its Theory and Application to Science, Technology and Humanities* (1979).

THOMAS F. GIERYN (EDITOR) was born in Rochester, New York (1950), and received the B.A. from Kalamazoo College (1972), and the M.A. (1975) and the Ph. D. (1979) from Columbia University. He is Assistant Professor of Sociology at Indiana University, Bloomington, and serves on the editorial board of *The American Sociologist*. His publications include "Problem Change and Problem Retention in Science" in *Sociological Inquiry* (1979).

ADOLF GRÜNBAUM is the Andrew Mellon Professor of Philosophy, Chairman of the Center for Philosophy of Science, and Research Professor of Psychiatry, University of Pittsburgh. He was born in Cologne (1923), and received the B.A. from Wesleyan University (1943), the M.S. in physics (1948) and the Ph. D. in philosophy (1951) from Yale University. A Fellow of the American Academy of Arts and Sciences, he has been appointed an Einstein Centennial Lecturer under the auspices of the Institute for Advanced Study, Princeton. His books include *Philosophical Problems of Space and Time* (enlarged edition, 1973) and *Modern Science and Zeno's Paradoxes* (revised edition, 1968).

IRVING JANIS has been teaching at Yale University since 1947, where he is now Professor of Psychology. Born in Buffalo, New York (1918), he received the B.S. from the University of Chicago (1939) and the Ph.D. from Columbia University (1948). He has received a Guggenheim Foundation Fellowship, and has been a fellow at the Center for Advanced Study in the Behavioral Sciences (1973–74). His publications include *Decision Making: A Psychological Analysis of Conflict, Choice and Commitment* (with L. Mann, 1977) and *Victims of Groupthink: A Psychological Study of Foreign-Policy Decisions and Fiascoes* (1972).

MICHAEL J. MULKAY, Reader in Sociology at the University of York, was born in London (1936). He received the B.A. from the London School of Economics (1965) and the Ph. D. from the University of Aberdeen (1970). His books include *Functionalism, Exchange and Theoretical Strategy* (1971), *Astronomy Transformed* (with David O. Edge, 1976), and *Science and the Sociology of Knowledge* (1979).

PAUL A. SAMUELSON, born in Gary, Indiana (1915), is Institute Professor at M.I.T., where he has been teaching economics since 1940. He received the B.A. from the University of Chicago, and the M.A. and the Ph.D. (1941) from Harvard University. He has been awarded the Nobel Prize in Economic Sciences (1970) and honorary degrees from a number of universities including Chicago (1961), Indiana (1966), Michigan (1967) and Harvard (1972). His textbook *Economics* is about to appear in its eleventh edition; his 1947 *Foundations of Economic Analysis* has been widely translated; and his pre-1977 *Collected Scientific Papers* have been published in four volumes.

GEORGE J. STIGLER was born in Renton, Washington (1911). He is the Charles R. Walgreen Distinguished Service Professor of American Institutions at the University of Chicago. He received the B.B.A. from the University of Washington (1931), the M.B.A. from Northwestern University (1932), and the Ph.D. from the University of Chicago (1938). He is a member of the National Academy of Sciences, and he has served as President of the Council of the American Economic Association (1964). His publications include *The Organization of Industry* (1968), *Essays in the History of Economics* (1965) and *The Citizen and the State* (1975).

STEPHEN STIGLER is Professor in the Department of Statistics, University of Chicago. Born in Minneapolis (1941), he received the B.A. from Carleton College and the Ph. D. from the University of California, Berkeley. He has previously taught at the University of Wisconsin, and has just assumed editorship of the *Journal of the American Statistical Association: Theory and Methods* (1979). He has been a Guggenheim Fellow and a Fellow at the Center for Advanced Study in the Behavioral Sciences. His publications include "Laplace's Early Work: Chronology and Citations" in *ISIS* (1978), and "Francis Ysidro Edgeworth, Statistician" in *J. Royal Statistical Society A* (1978).

Name Index

Adler, Alfred, 79–80
Adrain, Robert, 155
Aftalian, Albert, 128
Allen, F. L., 144–45
Anderson, C. G., 110
Anderson, Jack, 83, 88
Anderson, James, 145
Arendt, Hannah, 21, 27
Aristotle, 18–20, 27
Atthowe, J., 92, 108
Axelrad, Sidney, 69

Bacon, Francis, 35, 39, 42–43, 48, 59
Bales, V., 101
Barber, Elinor, 70, 72, 104, 143
Barone, Enrico, 143, 145
Bateson, William, 130
Beaver, Donald DeB., 155
Ben-David, Joseph, 60, 124
Becquerel, A. H., 2, 6, 14
Bar-Hillel, Y. 41–42
Benjamin, Walter, 34, 41
Berger, Gaston, 1
Bergman, A. B., 108
Berkowitz, Leonard, 69
Bernard, Claude, 1, 2, 4–6, 10–14
Bernardelli, H. 137
Bernoulli, Daniel, 127
Bernoulli, James, 126
Bernoulli, John, 126
Berscheid, E., 95, 98, 108, 110
Bertrand, J., 152
Bienaymé, Alain, 148, 150
Blake, William, 44, 48, 55, 60
Blaug, Mark, 146
Bliss, G. A., 127
Bloch, E., 20, 27
Böckh, August, 131
Bogdonoff, M. D., 92, 110
Bolyai, Wolfgang, 127
Borel, E. 131
Bottome, P., 79
Boulding, Kenneth, 137, 142
Bowers, K.G., 107–08
Box, Joan Fisher, 141–42
Box, Steven, 124
Brecht, B., 34
Brenner, M. H., 74
Broom, Leonard, 70, 72
Brown, E. S., 155
Brownell, K. D., 108
Buber, Martin, 77

Butler, Nicholas Murray, 95
Bynum, R., 108
Byme, D., 98, 108

Camus, Albert, 18–19, 27
Carnap, R., 41
Carter, Jimmy, 83, 88
Cartwright, D., 108
Cassel, Gustav, 127, 129
Cauchy, Augustin, 148
Chamberlin, E. H., 144–46
Chang, P., 100, 108
Chamy, E., 92, 108
Chauveau, Auguste, 12–13
Cicero, 26
Cioffi, Frank, 84–87, 89
Clark, John Bates, 127-28, 143
Clark, John Maurice, 128
Clinard, Marshall, 70, 72
Cloward, Richard, 69
Cohen, Albert K., 69
Colby, K. M., 79, 90
Cole, Jonathan R., 73–74, 155
Cole, Stephen, 67, 73–74, 154–55
Coleridge, S. T., 129
Colten, M. E., 101, 108
Conolley, E., 99–100, 108–09
Cope, Jackson I., 58
Copernicus, 76
Coser, Lewis A., 74, 110
Cotgrove, Stephen, 124
Cottrell, L. S., Jr., 70, 72
Cotz, R. E., 74
Cournand, André, 13, 60, 124
Courrier, Robert, 5
Cressey, Donald, 69
Cronbach, Lee J., 60
Crowley, Kate, 141–42
Crozier, Michel, 59
Curelaru, M., 28

Darwin, Charles, 76
Das, Man Singh, 110
daVinci, Leonardo, 85
Davis, M. S., 92–93, 108–09
De Moivre, Abraham, 151
Dennis, Wayne, 71–72
Descartes, R., 43, 81
Detienne, M., 41
Dewey, John, 45–46, 58–59
Dittes, J. E., 95, 109

163

Subject Index